STUDENT UNIT GUIDE

# A2 Human Biology
# UNIT 9a

# AQA

Specification A

Module 9a: Synoptic Assessment

Steve Potter

A2 Human Biology

Philip Allan Updates
Market Place
Deddington
Oxfordshire
OX15 0SE

tel: 01869 338652
fax: 01869 337590
e-mail: sales@philipallan.co.uk
www.philipallan.co.uk

© Philip Allan Updates 2005

ISBN 0 86003 943 9

All rights reserved; no part of this publication may be reproduced, stored in a retrieval system, or transmitted, in any form or by any means, electronic, mechanical, photocopying, recording or otherwise without either the prior written permission of Philip Allan Updates or a licence permitting restricted copying in the United Kingdom issued by the Copyright Licensing Agency Ltd, 90 Tottenham Court Road, London W1P 9HE.

This Guide has been written specifically to support students preparing for the AQA Specification A A2 Human Biology Unit 9a examination. The content has been neither approved nor endorsed by AQA and remains the sole responsibility of the author.

Printed by Information Press, Eynsham, Oxford

# Contents

## Introduction
About this guide ................................................................................................ 4
Preparing for the Unit 9a test ........................................................................ 4
Approaching the unit test .............................................................................. 5

■ ■ ■

## Content Guidance
About this section ............................................................................................ 8
**Module 1:** Molecules, Cells and Systems ................................................ 9
**Module 3:** Pathogens and Disease ........................................................ 18
**Module 5:** Inheritance, Evolution and Ecosystems .............................. 31
**Module 7:** The Human Life Span ............................................................ 49
Thinking synoptically .................................................................................... 78

■ ■ ■

## Questions and Answers
About this section .......................................................................................... 88
**Q1** Structured question .............................................................................. 89
**Q2** Comprehension question .................................................................... 93
**Q3** Essay question ...................................................................................... 97

# Introduction

## About this guide

This guide is written to help you to prepare for the Unit 9a examination of the AQA Human Biology Specification A. Unit 9a is a synoptic paper and forms part of the A2 assessment.

This Introduction provides guidance on revision, together with advice on approaching the examination itself.

The Content Guidance section summarises the whole specification. It does not go into full detail about every fact and you should *not* use this book to prepare for individual unit tests. It is assumed that you are already familiar with the content of individual modules. Where you are in doubt, you should refer to the relevant guide in this series, or other resources. The main principles covered in each module are explained and ways in which the principles of one section are linked to processes in another are outlined. There are suggestions about the types of question you might be asked. By highlighting many of the synoptic links for you, I hope that you will see patterns in the material and start to identify other links for yourself. The 'thinking synoptically' section explains what is expected of you in synoptic questions. Some strategies are suggested to enable you to give truly synoptic answers. This is of particular relevance to the essay questions.

The Question and Answer section shows you the sort of questions you can expect in the unit test. It would be impossible to give examples of every kind of question in one book, but these should give you a flavour of what to expect. Each question has been attempted by two candidates, Candidate A and Candidate B. Their answers, along with the examiner's comments, should help you to see what you need to do to score a good mark — and how you can easily *not* score a mark even though you probably understand the biology. In the essay questions, the notes also show how the examiner is thinking about the structure of the essay, as well as the content.

## Preparing for the Unit 9a test

Preparation for examinations is a very personal thing. Different people prepare, equally successfully, in very different ways. The key is being totally honest about what actually works for *you*. This is *not* necessarily the same as the style you would like to adopt. It is no use preparing to a background of rock music if this distracts you.

Whatever your style, you must have a plan. Sitting down the night before the examination with a file full of notes and a textbook does not constitute a revision plan — it is just desperation — and you must not expect a great deal from it. Whatever your personal style, there are a number of things you *must* do and a number of other things you *could* do.

# AQA(A) Unit 9a

## Things you *must* do

- Leave yourself enough time to cover *all* the material.
- Make sure that you actually *have* all the material to hand (use this book as a basis).
- Identify weaknesses early in your preparation so that you have time to do something about them.
- Make sure that you do something about these weaknesses.
- Familiarise yourself with the terminology used in examination questions (see below).

## Things you *could* do to help you learn

- Write a precis of portions of your notes, including all the relevant key points.
- Produce unlabelled diagrams of relevant structures, photocopy them and practise labelling them.
- Practise creating flow charts that summarise processes.
- Write key points on postcards (carry them around with you for a quick revise during a coffee break!).
- Discuss a topic with a friend also studying the same course.
- Try to explain a topic to someone *not* on the course.
- Practise examination questions on the topic.

# Approaching the unit test

## Terms used in examination questions

You will be asked precise questions in the examinations, so you can save a lot of valuable time as well as ensuring you score as many marks as possible by knowing what is expected. Terms most commonly used are explained below.

### Describe
This means exactly what it says — 'tell me about...' — and you should not need to explain why.

### Explain
Here you must give biological reasons for *why* or *how* something is happening.

### Complete
You must finish off a diagram, graph, flow chart or table.

### Draw/plot
This means that you must construct some type of graph. For this, make sure that you:
- choose a scale that makes good use of the graph paper (if a scale is not given) and does not leave all the plots tucked away in one corner
- plot an appropriate type of graph — if both variables are continuous variables, then a line graph is usually the most appropriate; if one is a discrete variable, then a bar chart is appropriate
- plot carefully using a sharp pencil and draw lines accurately

# A2 Human Biology

**From the...**
This means that you must use only information in the diagram/graph/photograph or other forms of data.

**Name**
This asks you to give the name of a structure/molecule/organism etc.

**Suggest**
This means 'give a plausible biological explanation for' — it is often used when testing understanding of concepts in an unfamiliar situation.

**Compare**
In this case you have to give similarities *and* differences between...

**Calculate**
This means add, subtract, multiply, divide (do some kind of sum!) and show how you got your answer — *always show your working!*

When you finally open the test paper, it can be quite a stressful moment. You might not recognise the diagram or graph used in question 1. It can be quite demoralising to attempt a question at the start of an examination if you are not feeling very confident about it. So:

- *do not* begin to write as soon as you open the paper
- *do not* answer question 1 first, just because it is printed first (the examiner did not sequence the questions with your particular favourites in mind)
- *do* scan *all* the questions before you begin to answer any
- *do* identify those questions about which you feel most confident
- *do answer first* those questions about which you feel most confident, regardless of order in the paper
- *do read the question carefully* — if you are asked to explain, then explain, don't just describe
- *do* take notice of the mark allocation and don't supply the examiner with all your knowledge of osmosis if there is only 1 mark allocated (similarly, you will have to come up with four ideas if 4 marks are allocated)
- *do* try to stick to the point in your answer (it is easy to stray into related areas that will not score marks and will use up valuable time)
- *do* take care with
  - drawings — you will not be asked to produce complex diagrams, but those you do produce must resemble the subject
  - labelling — label lines *must touch* the part you are required to identify; if they stop short or pass through the part, you will lose marks
  - graphs — draw *small* points if you are asked to plot a graph and join the points with ruled lines or, if specifically asked for, a line or smooth curve of best fit through all the points
- *do try* to answer *all* the questions

# Content Guidance

A2 Human Biology

This section is a guide to **Module 9a: Synoptic Assessment**. It contains a brief summary of the main topics covered by each of Modules 1, 3, 5 and 7. The aim is to remind you of the main principles and concepts of each module and to show how these can be linked to other areas of the specification in synoptic questions.

### Module 1: Molecules, Cells and Systems

Module 1 deals with basic biochemistry and cell biology and is the basis for the other modules. It is relevant to topics such as photosynthesis, digestion, respiration, excretion and control by nerves and hormones. In addition, it covers circulation and breathing in mammals. These topics are often linked to gas exchange, digestion and excretion.

### Module 3: Pathogens and Disease

Module 3 examines the means by which pathogenic organisms enter the body and cause disease, and how the body responds to infection. The module covers genetic encoding of information and genetic control of protein synthesis, and reviews the ways in which gene technology can be used in the development of products used to treat diseases. Consideration is also given to the impact on health of non-communicable diseases.

### Module 5: Inheritance, Evolution and Ecosystems

Module 5 introduces the ideas of Mendelian inheritance and how the behaviour of genes is linked to the behaviour of chromosomes. Natural selection and speciation are the basis for the appearance of new species and evolution. The flow of energy through ecosystems is linked to cycling of nutrients and the biochemistry of photosynthesis and respiration. Methods of studying ecosystems are included, while some of the influences of people on these ecosystems are considered.

### Module 7: The Human Life Span

Module 7 covers important aspects of reproduction, growth, development and ageing. It also examines the physiological processes of digestion, integration by nerves and hormones, energy transduction by sense cells and the principles of homeostasis.

AQA(A) Unit 9a

# Module 1: Molecules, Cells and Systems

## Microscopy

Microscopes produce magnified images of specimens.

**Magnification** is how much bigger the image appears.

$$\text{magnification} = \frac{\text{apparent size}}{\text{real size}}$$

Both sizes must be in the same units.

**Resolution** describes how well the microscope can distinguish between two points that are close together. Visual acuity is a similar concept.

Light microscopes, transmission electron microscopes and scanning electron microscopes produce different kinds of image.

| Feature | Light microscope | Transmission electron microscope | Scanning electron microscope |
| --- | --- | --- | --- |
| Illumination | Light | Electrons | Electrons |
| Resolution | 0.2 μm | 0.001 μm | 0.01 μm |
| Specimens | Living or dead whole cells or small organisms; may be stained | Dead sections of cells; dried and stained | Dead; can be whole cells or small organisms; may be stained |
| Image | Large organelles visible; no internal organelle detail | All organelles visible; internal structures shown; may have artefacts | 3-D images of whole structures |

**Links** You might be asked to:
- calculate the magnification, or the real or apparent size of any organ or tissue, from a photograph or drawing of a photograph
- identify organelles from photographs and suggest explanations for their presence or abundance
- explain whether a photograph was taken using a light microscope or an electron microscope

*Tip* The resolution of the image will help you to decide on the type of microscope used.

## Cell structure

You need to be able to recognise both **prokaryotic** (bacterial) and **eukaryotic** (plant and animal) cells.

| Feature | Prokaryotic cell | Plant cell | Animal cell |
|---|---|---|---|
| Nucleus | Absent | Present | Present |
| DNA | Circular 'Naked' | Linear; in chromosomes, with protein | Linear; in chromosomes, with protein |
| Ribosomes | Small | Large | Large |
| Mitochondria | Absent | Present | Present |
| Chloroplasts | Absent | Present | Absent |
| Lysosomes | Absent | Absent | Present |
| Cell wall | Non-cellulose | Cellulose | Absent |
| Capsule | Often present | Absent | Absent |
| Cell membrane | Present | Present | Present |
| Flagellae | Often present | Absent | Generally absent |

**Links** You might be asked to relate the abundance of these organelles in a cell to its function.

*Tip* Cells with many mitochondria need a lot of energy. Those with many ribosomes make a lot of protein. Secretory cells often have extensive Golgi bodies. Absorbing cells often have microvilli and many mitochondria (e.g. cells in intestine and nephron).

## Transport across membranes

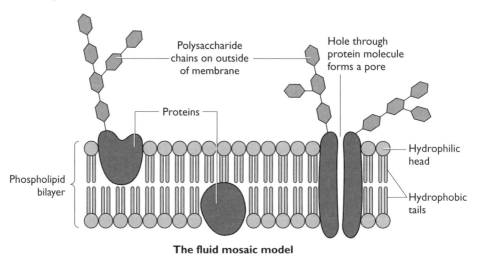

**The fluid mosaic model**

Substances move across plasma membranes by simple diffusion, facilitated diffusion, osmosis or active transport.

You must be able to relate uptake by carrier proteins or ion channels to:
- the tertiary shape of the protein
- the nature of the substance being transported

AQA(A) Unit 9a

|  | Simple diffusion | Facilitated diffusion | Osmosis | Active transport |
|---|---|---|---|---|
| Movement of molecules | Down a concentration gradient | Down a concentration gradient | Down a water potential gradient | Against a concentration gradient |
| ATP required | No | No | No | Yes |
| Carrier protein required | No | Yes | No | Yes |
| Substances moved | Small, non-polar | Specific to carrier protein | Water | Specific to carrier protein |

**Links** You might be asked to identify a process as simple diffusion, facilitated diffusion, osmosis or active transport.

*Tip* Clues in diagrams and data include:
- many mitochondria (ATP for active transport)
- process stops if respiration is not possible (active transport)
- evidence of water potential gradient (osmosis)
- carrier proteins (active transport or facilitated diffusion)

**Fick's law** summarises the factors affecting diffusion rate *across an exchange surface* (e.g. plasma membrane, wall of alveolus or wall of gill lamellae).

$$\text{rate of diffusion} \propto \frac{\text{concentration difference} \times \text{surface area}}{\text{thickness of exchange surface}}$$

All the processes are affected by temperature because molecules with more kinetic energy move faster. Facilitated diffusion and active transport are also dependent on the number of transport proteins per $\mu m^2$.

Concentration gradients are often maintained by removing the substance as soon as it moves across the surface (e.g. by blood or by active transport out of a cell). Circulatory systems are often important in maintaining concentration gradients.

**Links** You might have to use Fick's law to explain why an exchange surface is effective.

*Tip* Relate the three concepts in Fick's law to actual features of the structure you are describing (e.g. microvilli have a large surface area).

In addition to these processes, **exocytosis** moves large molecules *out of* cells; **endocytosis** moves large molecules *into* cells.

## Biological molecules

### Carbohydrates

**Monosaccharides** are single sugars. **Pentoses** contain five carbon atoms; they occur in nucleic acids. **Hexoses** contain six carbon atoms. You need to know the structure of the hexose sugars **α-glucose** and **β-glucose**. α-Glucose is the most common respiratory substrate.

**Disaccharides** are formed by a **condensation** reaction between two monosaccharides. **Maltose** consists of two α-glucose molecule residues linked by a **1,4 glycosidic bond**.

**Polysaccharides** are **macromolecules**. They are **polymers** formed from many monosaccharides joined by glycosidic bonds. **Starch** (plants) and **glycogen** (animals) are formed by condensation reactions between hundreds of α-glucose molecules. **Cellulose** (plants) consists of hundreds of β-glucose molecules linked by condensation. Starch and glycogen are **osmotically inactive**, compact molecules, making them effective storage carbohydrates. The fibrous nature of cellulose gives strength to cell walls.

## Lipids
A **triglyceride** molecule consists of three **fatty acids** linked by **ester bonds** to a molecule of **glycerol**. The ester bonds are formed by condensation reactions. A **phospholipid** molecule has two fatty acids and one phosphate group linked to a molecule of **glycerol**.

## Proteins
**Proteins** are macromolecules. They are polymers consisting of chains of **amino acids** linked by **peptide bonds** formed by condensation reactions. The sequence of amino acids is the **primary structure** of the protein. The **secondary structure** is the α-helix into which the primary structure coils. It is held together by **hydrogen bonds**. The **tertiary structure** is the final shape of the molecule, held together by **disulphide** and **ionic bonds**.

The specific tertiary structure of each protein molecule allows it to carry out its function as an enzyme (e.g. amylase), structural protein (e.g. collagen, actin and myosin) or receptor site (e.g. for hormones).

## Hydrolysis
The bonds linking the subunits in macromolecules are broken by **hydrolysis** during **digestion** in animals and when food stores are mobilised in plants.

$$\begin{array}{c}
H \\
| \\
H-C-O-C(=O)-R^1 \\
| \\
H-C-O-C(=O)-R^2 \\
| \\
H-C-O-C(=O)-R^3 \\
| \\
H
\end{array} + 3H_2O \underset{\text{condensation}}{\overset{\text{hydrolysis}}{\rightleftharpoons}} \begin{array}{c}
H \\
| \\
H-C-OH \\
| \\
H-C-OH \\
| \\
H-C-OH \\
| \\
H
\end{array} + \begin{array}{c}
O \\
|| \\
HO-C-R^1 \\
O \\
|| \\
HO-C-R^2 \\
O \\
|| \\
HO-C-R^3
\end{array}$$

Ester bond

## Identification
The different molecules are identified by biochemical tests.

AQA (A) Unit 9a

| Test | Reagent | Reagent colour | Method | Positive result |
|---|---|---|---|---|
| Starch | Iodine solution | Yellow | Add iodine solution to substance | Blue-black |
| Reducing sugar | Benedict's reagent | Blue | Heat substance with Benedict's reagent | Yellow/orange/red |
| *Non-reducing sugar | Benedict's reagent | Blue | Hydrolyse by boiling with HCl; neutralise with $NaHCO_3$; heat with Benedict's reagent | Yellow/orange/red |
| Protein | Biuret solution | Blue | Add Biuret solution to substance | Mauve/lilac/purple |
| Lipid | Alcohol and water | Colourless | Mix substance with alcohol; filter into test tube of water | Cloudy/milky |

\* Only carry out the non-reducing sugar test after a reducing sugar test has proved negative

## Enzymes

Enzymes control all metabolic processes (e.g. respiration, photosynthesis, protein synthesis and DNA replication), so factors that affect enzymes affect these processes.

All enzymes are:
- globular proteins with a specific tertiary structure, including a precisely shaped **active site**
- **catalysts** that speed up reactions by lowering **activation energy**
- **specific**, because the shape of the active site allows only one substrate to bind and form an **enzyme–substrate complex**
- affected by pH and temperature, enzyme concentration, substrate concentration and inhibitors

*Tip* Make sure that you use the term 'active site' only when you are writing about enzymes.

**Optimum** conditions allow enzymes to function most effectively. High temperatures and extremes of pH **denature** enzymes, altering the structure of their active sites.
- **Competitive inhibitors** are molecules that bind to the active site of an enzyme, thus excluding substrate molecules. The extent of inhibition depends on the ratio of inhibitor to substrate.
- **Non-competitive inhibitors** bind to **allosteric sites** and distort the active site so that the substrate cannot bind. The extent of inhibition depends solely on the amount of inhibitor.

# A2 Human Biology

## Cells, tissues and organs

Groups of similar cells organised to perform similar functions are called **tissues**. Usually there is just one type of cell in a tissue. Blood is unusual because it contains several types of cell and the tissue is liquid.

Other examples of tissue include:
- epithelia (singular epithelium)
- connective tissue
- nervous tissue
- xylem

*Tip* Make sure that you use the term 'cell' accurately. Mitochondria, chloroplasts and other cell structures are organelles. Alveoli are multicellular structures, not cells.

**Links** You might have to relate the structure of tissues to their function:
- Squamous epithelial cells in the alveoli are thin to give a short diffusion pathway.
- Columnar epithelial cells in the ileum have microvilli to increase the area for absorption.

**Organs** are structures that contain several tissues. Each tissue makes a specific contribution to the overall function of the organ.

The heart contains:
- cardiac muscle — to generate the force to pump the blood
- blood in the coronary vessels — to transport oxygen and nutrients to cardiac muscle and to remove excretory products
- Purkyne tissue — to transmit impulses that initiate contraction of the cardiac muscle

These are all necessary for the heart to beat with the right force and at the correct rate.

**Links** You might have to explain why one structure is an organ when another is not.

*Tip* Remember, an organ contains more than one tissue. Arteries and veins are organs because they contain muscle tissue, fibrous tissue and endothelium. A capillary contains only endothelium and so is not an organ. A muscle is an organ (it contains muscle tissue, nervous tissue and blood) and so is a nerve (it contains nervous and connective tissues).

## The heart and circulation

Blood is circulated by contractions of the ventricle walls. The output per ventricle *per beat* is the **stroke volume** and the number of beats per minute is the **heart rate**. The output *per ventricle per minute* is the **cardiac output**.

**Links** You might have to interpret data on the adaptations of the circulatory systems of mammals. For example:
- Small mammals readily lose heat because they have a high surface area to volume ratio. Therefore, their heart rate is high, in order to distribute heat effectively.
- Some mammals (e.g. people living at high altitudes and diving mammals) have to absorb oxygen very efficiently. Their blood often has more red blood cells per $cm^3$.
- Patterns of circulation may alter to meet unusual circumstances.

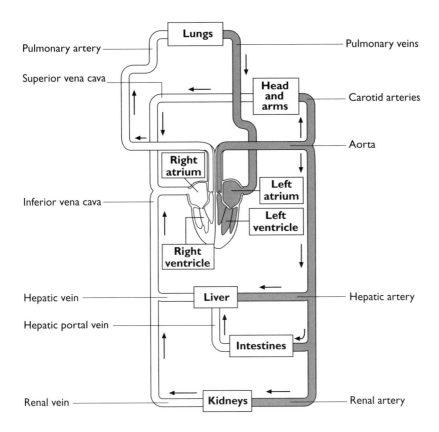

Although the heart *beat* is **myogenic**, heart *rate* is controlled by the **cardiac centre** in the **medulla** which sends impulses along **neurones** of the **sympathetic** and **parasympathetic** branches of the **autonomic nervous system** to the **sinoatrial (SA) node**. The neurones release different neurotransmitters that bind to specific receptor sites in the SA node. Sympathetic neurones release **noradrenaline**, which has an excitatory effect on the SA node. Parasympathetic neurones release **acetylcholine**, which has an inhibitory effect on the SA node.

**Links** You might have to use this example to explain the **antagonistic action** of the two branches of the autonomic nervous system. Drugs that block the action of either branch often have molecular shapes similar to the **neurotransmitters** and complementary to the **receptor sites** in the SA node.

### Blood vessels

Arteries and arterioles carry high-pressure blood quickly to organs. Capillaries carry blood through organs. Venules and veins carry low-pressure blood back towards the heart. Their structures are related to their functions.

- **Arteries** have large amounts of smooth muscle and elastic tissue in their walls to allow stretching under high pressure, followed by recoil. They act as secondary pumps.

- **Arterioles** have proportionately more smooth muscle in their walls to allow constriction and dilation. This mechanism allows the blood supply to an organ to be altered.
- **Capillaries** are very small. They are close to all cells in an organ and have walls that are only one cell thick. This allows rapid exchange of materials. Their walls are 'leaky' and tissue fluid escapes.
- **Venules** and **veins** have relatively thin walls because they do not have to withstand much pressure. Their large lumens minimise resistance to blood flow and valves prevent backflow. Blood is often moved through veins by the action of skeletal muscles surrounding the vein.

**Links** You might have to relate the structure and/or position of blood vessels to *specific* functions. For example, skin arterioles constrict or dilate to alter blood flow to the skin to vary heat loss from the skin.

***Tip*** If you are asked about the importance of a blood vessel in a certain situation, try substituting a description of the structure for the name in the question. For example, 'What is the importance of skin arterioles (*small arteries with walls containing smooth muscle*) in temperature control?'

## The lungs and breathing

You should understand the link between breathing and respiration. Breathing obtains the oxygen needed for aerobic respiration and removes the carbon dioxide produced in the process.

The amount of air breathed in per breath is the **tidal volume**. The **pulmonary ventilation** is the total volume of air breathed in per minute (tidal volume × breathing rate).

Alveolar structure is adapted for efficient gas exchange:
- The walls of alveoli are thin, so there is a short diffusion distance.
- Collectively, alveoli have a large surface area.

There is also a steep concentration gradient, maintained by ventilation and circulation.

***Tip*** If you are asked to use Fick's law to show how the structure of an exchange surface makes it effective, make sure that you describe the *actual structures*. For example, say *what* gives it the large surface area (the many alveoli) — don't just say it *has* a large surface area.

**Links** You need to be able to link ventilation, circulation and concentration gradients with the association of haemoglobin and oxygen to form oxyhaemoglobin. You need to be able to interpret haemoglobin dissociation curves.

***Tip*** Haemoglobin and oxygen only associate when there is a high partial pressure of oxygen. Ventilation and circulation ensure a steep diffusion gradient, so oxygen diffuses into red blood cells. Therefore, the partial pressure of oxygen in the red blood cells is high and association occurs.

### Control of breathing

Breathing is under the control of the autonomic nervous system. The **inspiratory centre** in the medulla responds to impulses from stretch receptors in the thorax. The basic control of breathing *at rest* is shown below.

- The inspiratory centre sends impulses causing intercostal and diaphragm muscles to contract. Inspiration begins, stimulating stretch receptors in the wall of the thorax.
- Stretch receptors begin to send inhibitory impulses to the inspiratory centre, which reduces the number of impulses to the intercostal and diaphragm muscles.
- Inspiratory centre inhibition increases and, eventually, no impulses are sent to the muscles and inspiration is inhibited.
- Expiration occurs passively and stretch receptors send fewer inhibitory impulses; at the end of expiration, inhibition of the inspiratory centre is removed.

**Links** You might be asked to use this to explain negative feedback. In this process, a change in conditions brings about measures that counteract the change and restore the original condition.

*Tip* In describing negative feedback you might describe:
- *the change* — impulses from the inspiratory centre initiate inspiration, which stretches the thorax
- *how the change is detected* — stretching is detected by stretch receptors
- *measures to counteract the change* — progressively more inhibitory impulses are sent to the medulla
- *how the original condition is restored* — inspiration ceases, expiration begins and the stretch is reduced.

## How exercise affects circulation and breathing

Exercise demands more activity by skeletal muscles. Therefore, these muscles must respire faster to release more energy. The breathing and circulatory systems must deliver more oxygen to the active muscles. During exercise, more oxygenated blood is delivered to muscles because:
- cardiac output increases (both stroke volume and heart rate increase)
- pulmonary ventilation increases (tidal volume and breathing rate increase)
- blood is diverted away from some organs to the skeletal muscles and skin

**Links** A question on exercise may bring in many aspects of changes to the circulatory and breathing systems, as well as aerobic and anaerobic respiration and transport of respiratory gases.

In active muscles there is increased dissociation of oxyhaemoglobin because pH is lowered by the increased carbon dioxide concentration — the **Bohr effect**. The increased dissociation of oxyhaemoglobin means more oxygen reaches active muscles. This results in increased aerobic respiration, which generates more ATP.

*Tip* Remember that the adaptations of the breathing and circulatory systems during exercise are aimed at:
- delivering more oxygenated blood to the muscles
- removing more carbon dioxide from the body
- maintaining a core temperature, if possible

# Module 3: Pathogens and Disease

## Pathogenic microorganisms

### The structure of pathogenic microorganisms
Most infectious diseases are caused by either bacteria or viruses.

**Structure of a typical bacterium**

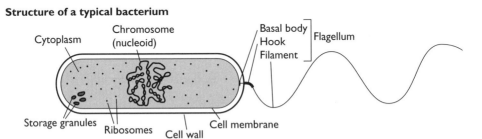

**Structure of a typical virus**

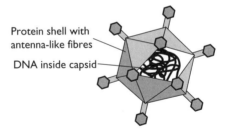

**Links** Bacterial cells are simple prokaryotic cells. You may have to cite the differences between these and eukaryotic cells, such as those of animals and plants.

### Reproduction in bacteria
Bacteria reproduce by **binary fission**. The DNA replicates and the cell divides into two. For optimum reproduction, bacteria require:
- optimum temperature for the action of bacterial enzymes
- optimum pH levels for the action of bacterial enzymes
- a supply of water — a bacterial cell is typically 80% water and cannot be active without it
- a supply of nutrients for synthesis of cellular structures and to release energy in respiration
- an adequate supply of oxygen, allowing aerobic respiration to release the energy necessary for reactions to proceed

The growth of a population of bacteria is shown in a **bacterial growth curve**.

# AQA (A) Unit 9a

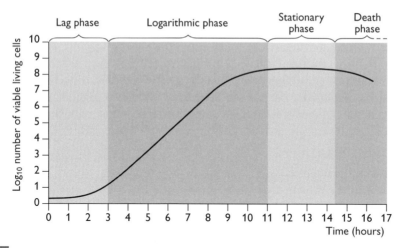

**Links** The rate of growth of a bacterial population could be linked to:
- respiration — the physiological processes involved in reproduction and growth require energy
- enzyme activity — the conditions in the culture (e.g. temperature, pH) may affect bacterial enzymes
- protein synthesis

## Counting numbers of bacteria in a liquid culture

The number of bacteria in a liquid culture can be estimated using a haemocytometer. This is a special type of microscope slide with a cavity of known depth precisely cut out and a measurement grid accurately engraved on its surface.

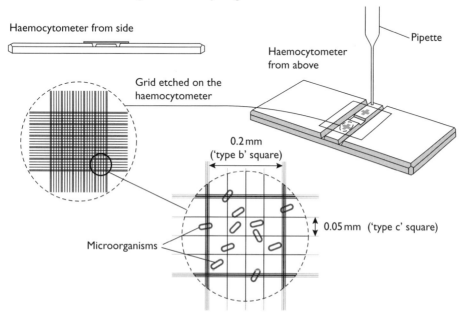

To use the haemocytometer, carry out the following:
- Place a sample of the culture under the cover slip by capillary action.
- Count the number of bacteria in several of the 'type b' squares as shown in the diagram. Some bacterial cells will be 'half in' and 'half out'. Record half of these bacteria as being 'in' the square.
- Obtain an average for the number of bacteria per type b square.
- Estimate the number of bacteria in 1 cm³ of culture in the following way.
  - each type b square has a side of 0.2 mm and therefore an area of

    $0.2 \times 0.2 = 0.04 \text{ mm}^2$

  - the depth of the cavity is 0.1 mm
  - therefore, the volume of a type b square = $0.04 \times 0.1 = 0.004 \text{ mm}^3$ or $0.000004 \text{ cm}^3$
  - if the mean number of bacteria per type b square was 28, the number of bacteria in 1 cm³ of the culture would be

    $$\frac{28}{0.000004} = 7\,000\,000$$

## The link between pathogenic microorganisms and disease

### Koch's postulates
Robert Koch suggested the following 'postulates' as being necessary to prove that a particular microorganism causes a particular disease. They state:
- the microorganism is always present when the disease is present and is absent if the disease is absent
- the microorganism can be isolated from an infected person and then grown in culture
- introducing such cultured microorganisms into a healthy host results in the disease developing
- the microorganism can then be isolated from this newly diseased host and grown in culture

### How bacteria and viruses cause disease
Bacteria are cellular and do not normally invade our cells following infection. However, they do produce toxins as a result of their metabolism and it is these toxins that cause the damage to the body. Viruses invade our cells and, once inside, direct the cell's metabolism to produce more viruses. This results in cell death, which is the cause of the disease.

### How diseases are transmitted

| Method of infection | Transmission of infection | Diseases spread in this way |
|---|---|---|
| Droplet infection | When an infected person sneezes, microorganisms carried in tiny droplets are breathed in by another person | Influenza, common cold, pneumonia |
| Drinking contaminated water | Microorganisms in the water infect cells lining the gut and then reproduce; they are later released back into the gut and pass out with the faeces | Cholera, typhoid fever |

| Method of infection | Transmission of infection | Diseases spread in this way |
|---|---|---|
| Eating contaminated food | Cells lining the gut are infected — microorganisms reproduce and are passed out with the faeces | Salmonellosis, botulism, listeriosis |
| Direct contact | Contact with an infected person's skin or with a contaminated surface | Athlete's foot, ringworm |
| Sexual intercourse | Microorganisms infecting the sex organs can be transmitted during intercourse | Syphilis, AIDS, gonorrhoea |
| Blood to blood contact | Many sexually transmitted diseases are transmitted this way; drug users sharing needles are also at risk | AIDS, hepatitis B |
| Animal vectors | Many diseases are spread through the bites of insects, often as they suck blood | Malaria, sleeping sickness |

## Specific diseases

| Disease | Microorganism | Transmission | Symptoms | How caused |
|---|---|---|---|---|
| Salmonellosis | *Salmonella* (bacterium) | Contaminated food | • Diarrhoea<br>• Fever | • Toxins prevent uptake of sodium and glucose from gut<br>• Water lost from cells by osmosis |
| Pulmonary tuberculosis (TB) | *Mycobacterium tuberculosis* (bacterium) | Droplet infection | • Coughing (blood)<br>• Shortness of breath | • Reproduction of bacteria damages lung tissue<br>• Damage to lung tissue reduces oxygen uptake |
| Acquired immune deficiency syndrome (AIDS) | Human immuno-deficiency virus (HIV) | Sexual intercourse | • HIV antibodies in blood<br>• Kapsosi's sarcoma<br>• A range of infections | • Response of immune system to infection by HIV<br>• Reduced immune efficiency as the virus destroys helper T-lymphocytes |

- Salmonellosis can be prevented by:
  - effective hygiene when preparing animal foods
  - effective hygiene when preparing human food
  - cooking food thoroughly
- Tuberculosis is most common in groups with reduced immune efficiency, which is often due to a low standard of living, for example a poor diet and inadequate housing.
- The risk of AIDS can be reduced by avoiding intercourse with high-risk groups of people or by using condoms. It is also important to discourage drug users from

sharing needles. As yet, there is no cure for AIDS, although certain treatments may help some people to manage the disease.

The structure of HIV and its life cycle are shown below.

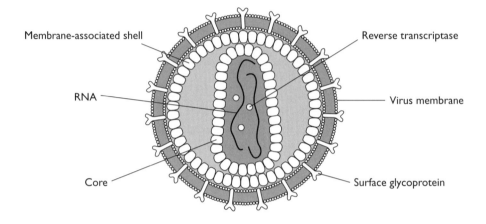

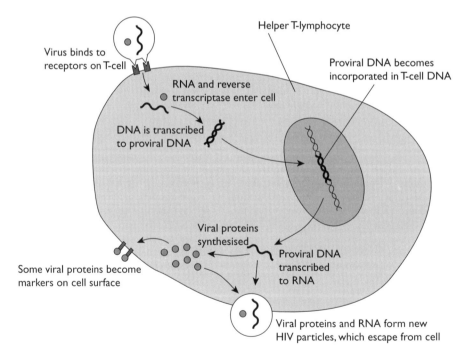

**Links** You may have to link the HIV life cycle to:
- protein synthesis (transcription of DNA to mRNA and translation of the mRNA)
- the action of reverse transcriptase

# Mitosis, meiosis and the cell cycle

**Links** Questions involving epithelia in particular might lead into cell replacement and then into mitosis and the cell cycle.

| Stage of process | Sequence for mitosis | Sequence for meiosis |
| --- | --- | --- |
| Interphase | Chromosomes duplicate | Chromosomes duplicate |
| Prophase (I) | Nuclear membrane breaks down<br><br>Chromosomes shorten and thicken; chromosomes consist of two chromatids | Nuclear membrane breaks down<br><br>Chromosomes shorten and thicken; chromosomes consist of two chromatids; homologous chromosomes pair up and crossing over occurs |
| Metaphase (I) (= middle) | Chromosomes align separately on spindle | Chromosomes align on spindle; homologous chromosomes are still in pairs |
| Anaphase (I) (= apart) | Spindle fibres pull a chromatid from each chromosome to opposite poles | Spindle fibres separate homologous chromosomes (independent assortment) |
| Telophase (I) | Two new nuclei form, each with the diploid number of single chromosomes | Two haploid nuclei form; chromosomes are double (two chromatids) |
| Prophase II | N/A | Chromosomes shorten |
| Metaphase II | N/A | Chromosomes in each cell align on spindle |
| Anaphase II | N/A | The chromatids from each chromosome are pulled to opposite poles |
| Telophase II | N/A | Four haploid nuclei form; chromosomes are single |
| Cytokinesis | Two diploid genetically identical cells | Four haploid cells showing genetic variation |

**Links** You might have to explain examples of Mendelian inheritance using knowledge of meiosis and random fertilisation.

*Tip* Remember that each symbol in a parental genotype represents a gene on a chromosome. In a typical life cycle:
- meiosis produces the haploid number of chromosomes
- fertilisation restores the diploid number
- mitosis maintains the diploid number

## The cell cycle

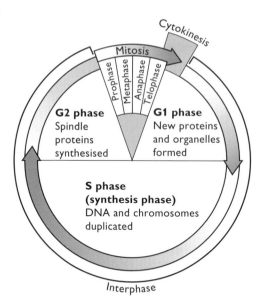

The G1 phase of the cell cycle involves protein synthesis; you might have to give some details of the process. The S phase involves DNA replication.

**Links** A question on life cycles in a synoptic paper might require details of crossing over and independent assortment as sources of variation, as well as knowing how meiosis helps to maintain a constant chromosome number from generation to generation.

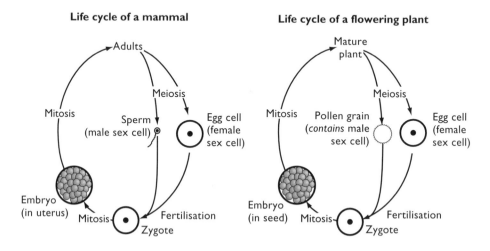

## The structure and function of DNA and RNA

You are unlikely to be asked simply to label a diagram of **DNA** or **mRNA** in a synoptic paper, but you should appreciate that structure is related to function.

| Feature | DNA | mRNA |
|---|---|---|
| Size of molecule | Large — carries the code for all proteins | Small — carries the code for one protein |
| Number of strands | Two — allows semi-conservative replication | One — codes for one protein and does not replicate |
| Stability of molecule | Very stable — changes in structure would alter the way the cell and daughter cells function | Unstable — if mRNA were stable, it could lead to over-production of the protein it codes for |

## DNA replication

A molecule of DNA consists of two anti-parallel strands. The key stages in DNA replication are shown in the diagram.

The polynucleotide strands of DNA separate as DNA helicase breaks the hydrogen bonds holding the strands together…

…each strand acts as a template for the formation of a new molecule of DNA…

…DNA polymerase assembles nucleotides into two new strands according to the base-pairing rule…

…two identical DNA molecules are formed — each contains a strand from the parent DNA and a new complementary strand

This is called **semi-conservative replication** because one original strand is retained in each new molecule of DNA. Faults in DNA replication give rise to **mutations**, including **insertions** and **deletions**.

## Protein synthesis

- DNA base triplets are **transcribed** to produce complementary triplets in mRNA, which are called **codons**. Each codon codes for a particular amino acid.
- In the ribosomes, mRNA codons are **translated** into amino acid sequences.
- **tRNA** molecules carry specific amino acids to the ribosome.
- Complementary **anticodons** on tRNA allow recognition of the codons on mRNA, producing the correct sequence of amino acids.

**Links** You might have to link mutation and/or protein synthesis with the functioning of a specific protein.

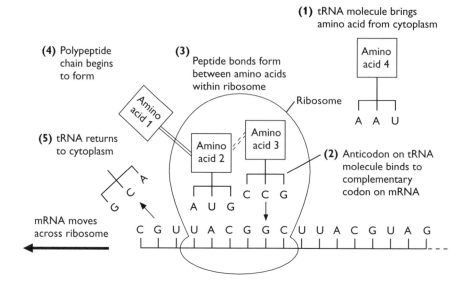

**Tip** Any change in the base sequence of DNA will also result in a change in the base sequence of mRNA. This will now code for a different sequence of amino acids, producing a protein with a different tertiary structure.

## Genetic engineering

Genes can be isolated for transfer to a bacterium either by 'cutting' the gene out using **restriction enzymes** to leave **sticky ends** on the gene or by manufacturing the gene from mRNA using the enzyme **reverse transcriptase**.

Genes are often transferred by **plasmids**, which are one kind of **vector**. The isolated genes are incorporated into the plasmids using **ligase enzymes**. The plasmids are cultured with bacteria, which take them up, so the bacteria have the gene to code for the product. Large-scale production involves growing bacteria in fermenters and then isolating the product.

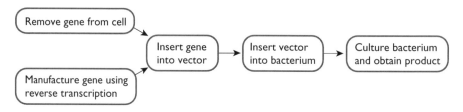

**Links** You might be asked (e.g. in a question on hormones) why a particular product of genetic engineering functions in the same way as the non-engineered substance.

**Tip** The product functions in exactly the same way as the non-engineered substance because it is coded for by the same gene (transferred during genetic engineering). The same mRNA is formed, which produces protein identical to the non-engineered substance.

## Diagnosis and treatment of disease

### Using DNA probes for the diagnosis of genetic diseases

Some diseases — such as cystic fibrosis, Huntington's chorea and some types of haemophilia — are genetically determined. DNA probes (also called gene probes) can be used to show that a particular gene is present in a sample of DNA. Each probe is a piece of single-stranded DNA with a base sequence complementary to that of the actual gene. They are either fluorescent or radioactive. DNA probes are used in the following way:

- A sample of DNA is extracted (usually from the nuclei of white blood cells, as blood is easily obtained).
- The DNA is heat-treated to split the double helix into single strands.
- The DNA is incubated with the DNA probe.
- After a time, any surplus probe is 'washed out'.
- The remaining material is checked for fluorescence (or radioactivity). If this is found, it must be because some probe is still present. This in turn must mean that the probe is bound to the DNA sample. Since it could only bind to its complementary DNA sequence, the gene in question must be present.

Genetic fingerprinting is also used to show if an individual carries a gene determining a specific disease.

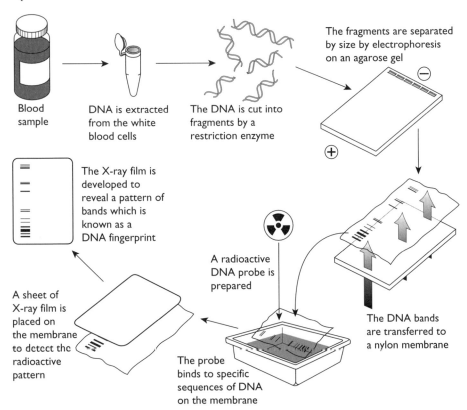

The so-called 'genetic fingerprint' can then be compared with one known to contain the specific disease-causing gene.

**Links** You may have to link genetic fingerprinting to the structure of DNA or to human reproduction and paternity issues.

### Using enzymes for diagnosis of disease

**Pancreatitis** is a condition in which the pancreas is inflamed and the pancreatic duct is blocked. As a result, pancreatic enzymes (amylase and lipase) cannot reach the duodenum. Instead, they pass into the blood plasma. An assay of the blood showing abnormally high concentrations of these enzymes is a strong indicator of pancreatitis.

**Diabetes** is a condition in which the cells of the Islets of Langerhans (in the pancreas) do not produce sufficient insulin. As a result, plasma glucose concentrations can become abnormally high and can rise above the **renal threshold** and so be excreted in the urine. The concentration of glucose in the urine can be checked using a paper strip in which is embedded the enzymes **glucose oxidase** and **peroxidase** together with a colourless dye. The paper strip is dipped in a sample of urine and the following reactions take place:

$$\text{glucose} + \text{oxygen} \xrightarrow{\text{glucose oxidase}} \text{gluconic acid} + \text{hydrogen peroxide}$$

$$\text{hydrogen peroxide} + \text{colourless dye} \xrightarrow{\text{peroxidase}} \text{coloured dye} + \text{water}$$

Different colours result from different concentrations of glucose in the urine.

## Treatment of disease

### Using antibiotics to treat disease

Antibiotics are used to treat disease caused by bacteria. They act by disrupting metabolic processes of bacterial cells, such as DNA replication, cell wall synthesis and protein synthesis. **Bactericidal** antibiotics kill bacteria. **Bacteriostatic** antibiotics slow down the rate of bacterial reproduction.

| Mode of action of antibiotic | Example | How the antibiotic works | Bactericidal or bacteriostatic? |
|---|---|---|---|
| Disrupts cell wall synthesis | Penicillin | Weakened cell wall cannot resist entry of water by osmosis and cell bursts (osmotic lysis) | Bactericidal |
| Disrupts DNA replication | Nalidixic acid | Bacteria are not killed, but cell division is halted | Bacteriostatic |
| Disrupts protein synthesis | Tetracycline | Bacterial cell cannot synthesise enzymes and structural proteins | Bactericidal |

The diagram below shows the mode of action of penicillin.

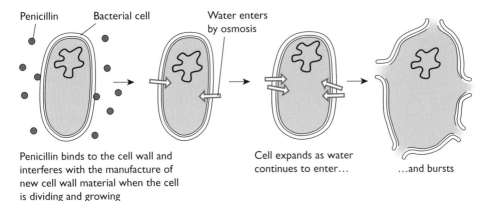

Penicillin binds to the cell wall and interferes with the manufacture of new cell wall material when the cell is dividing and growing

Cell expands as water continues to enter...

...and bursts

Antibiotics are ineffective against viruses. As viruses are acellular they do not carry out any of the processes targeted by antibiotics. Furthermore, viruses enter human cells; antibiotics cannot as there are no suitable transport proteins to carry them in.

**Links** You may have to link the action of antibiotics to:
- DNA replication
- protein synthesis
- the structure of cell walls and plasma membranes (with respect to the presence/absence of specific markers/receptors to which the antibiotics can bind)

### Using beta-blockers to treat hypertension

Hypertension is sustained high blood pressure and a key factor in coronary heart disease. It is caused by the continued over-secretion of the hormones **adrenaline** and **noradrenaline** at the endings of neurones of the **sympathetic division** of the **autonomic nervous system**, frequently as a result of **chronic stress**. These hormones bind to beta-receptors in the walls of the ventricles and cause them to contract with more force, raising the blood pressure. Beta-blocker molecules have shapes that are complementary to those of the beta-receptors. As a result, they bind to the receptors, preventing the hormones from binding.

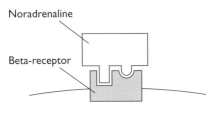

In the absence of beta-blockers, noradrenaline can bind to beta-receptors to increase heart rate and force of contraction

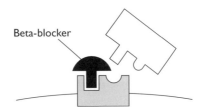

In the presence of beta-blockers, noradrenaline is prevented from binding to beta-receptors

**Links** You may have to link the action of beta-blockers to:
- the tertiary structure of protein molecules
- the control of the heartbeat by the autonomic nervous system

### Using monoclonal antibodies in diagnosis and treatment

Monoclonal antibodies are the product of a single type of plasma cell and will bind with only one specific antigen. Because of this precise binding, they can be used to detect antigens associated with:
- some cancer cells
- some viruses
- pregnancy

Pregnancy testing kits detect the presence of human chorionic gonadotrophin (hCG). This hormone is only produced by an implanted embryo. It passes into the maternal blood plasma and some is excreted in the urine. Presence of hCG in the urine therefore confirms pregnancy.

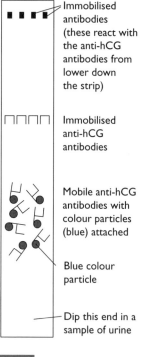

Immobilised antibodies (these react with the anti-hCG antibodies from lower down the strip)

Immobilised anti-hCG antibodies

Mobile anti-hCG antibodies with colour particles (blue) attached

Blue colour particle

Dip this end in a sample of urine

Urine rises up absorbent strip carrying hCG; this binds with anti-hCG antibodies, which are then also carried up the strip

Unbound anti-hCG antibodies bind here and a blue line is seen; this line will always form

Anti-hCG antibodies bound to hCG bind here and a blue line is seen; this blue line only forms if hCG is present and so confirms a pregnancy

**Links** You may have to link the use of monoclonal antibodies in pregnancy testing to related material on human reproduction and early development.

Monoclonal antibodies may be used to treat cancers by being bound to an anti-cancer drug. The antibody will then bind only with cells with the cancer antigen — cancer cells. This could reduce the dose of drugs needed to treat cancers, as the treatment is more selective.

# Module 5: Inheritance, Evolution and Ecosystems

## Meiosis

**Prophase I**
Diploid cell with two pairs of homologous chromosomes which have already duplicated
Nuclear membrane breaks down and crossing over occurs during this stage

Nucleus

**Metaphase I**
Homologous chromosomes align on spindle opposite each other

Spindle

**Anaphase I**
Independent assortment of homologous chromosomes — pulled apart by shortening of spindle fibres

**Telophase I**
Two haploid cells form
One chromosome from each homologous pair is found in each new nucleus

**Prophase II**
Each cell contains one chromosome from each homologous pair (no crossing over)

**Metaphase II**
Chromosomes align on spindle

**Anaphase II**
One chromatid from each chromosome is pulled to each pole

**Telophase II**
Four haploid nuclei, each with one chromosome from each homologous pair

Four haploid cells

(See the table on page 23.)

## Mendelian inheritance

**Genes** determine features. They often exist in different forms called **alleles**. Alleles are carried on homologous chromosomes. If an organism has two identical alleles for a feature (e.g. two blue alleles of the gene for eye colour) per cell, it is **homozygous** for that feature. If it has two different alleles per cell, it is **heterozygous** for that feature. Sometimes one allele is **dominant** over the other allele, which is **recessive**. A recessive allele is only expressed when the corresponding dominant allele is absent. The **genotype** of an organism is its genetic make up. The **phenotype** is what results from that genotype.

You should be able to use these terms correctly in any context, not just in questions about Mendelian inheritance.

**Monohybrid inheritance**

In monohybrid inheritance, the pattern of inheritance of the alleles of just one gene is studied. Different genes can show:
- dominant/recessive alleles (e.g. phenylthiocarbamide (PTC) tasting, albinism)
- codominant alleles (e.g. flower colour in antirrhinums)
- sex-linked inheritance (e.g. haemophilia, red–green colour blindness)
- multiple allele inheritance (e.g. ABO blood grouping — this also includes some codominance)

Full details of these types of inheritance are in the Module 5 guide in this series. The main principles of understanding monohybrid inheritance are:
- All cells, except the sex cells, contain two alleles of a gene on homologous chromosomes.
- Sex cells contain only one allele of a pair because they are formed by meiosis, which results in only one chromosome from each homologous pair being passed to daughter cells.
- Homozygous individuals produce only one kind of gamete because both alleles on the homologous chromosomes are the same.
- Heterozygous individuals produce two kinds of gamete because the homologous chromosomes carry different alleles and are separated in meiosis.
- Fertilisation is random — each possible fertilisation has an equal chance of taking place. The total number of possible fertilisations allows us to work out the probability that any one will occur.

*Tip* When solving problems, remember:
- each individual inherits one allele for a feature from each parent
- an individual showing the dominant feature must have inherited at least one dominant allele
- an individual showing the recessive feature must be homozygous
- an individual showing the recessive feature has inherited one recessive allele from each parent and will pass on one recessive allele to all children

To deduce whether an allele is dominant or recessive, look for parents showing the same feature (e.g. both can taste PTC) producing a child showing the alternative feature (non-taster). The allele for the alternative feature is recessive. In this example, the non-tasting alleles must have been inherited from the tasting parents. They must be heterozygous, so the tasting allele is dominant and the non-tasting allele is recessive.

*Codominance*

Codominant alleles are inherited in exactly the same way as alleles that are either dominant or recessive. In codominance, however, heterozygous individuals show a condition that is influenced by *both alleles*.

In antirrhinums (snapdragons), flower colour is determined by two codominant alleles, **R** (for red flowers) and **r** (for white flowers).

- **RR** — homozygote with red flowers
- **rr** — homozygote with white flowers
- **Rr** — heterozygote with pink flowers

*Tip* Remember, a cross involving codominance between two heterozygotes produces a genotype ratio of 1:2:1 and a phenotype ratio of 1:2:1. A cross involving codominance between a heterozygote and a homozygous recessive produces a genotype ratio of 1:1 and a phenotype ratio of 1:1.

*Sex linkage*
Some conditions are determined by genes carried on the sex chromosomes. They are **sex-linked**. Examples include some types of haemophilia and red–green colour blindness.

Questions about sex linkage are often based on information presented as **pedigrees**.

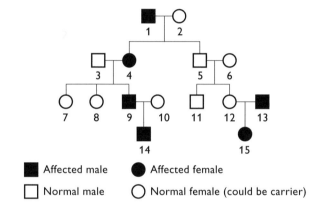

These conditions are determined by recessive alleles carried on the X chromosome and are therefore more common in males than in females. Males only have one X chromosome and if that chromosome carries the recessive allele they will show the condition. Females have two X chromosomes and so both must carry the recessive allele if they are to show the condition.

**Links** You might be asked to decide whether a feature is sex-linked. This is difficult to prove, but there are some strong indicators:
- The feature is more common in males than females. In the pedigree above, four out of seven males are affected, but only two out of eight females are affected.
- No affected female has an unaffected male parent.
- It sometimes skips generations.

## Dihybrid inheritance
In dihybrid inheritance, the patterns of inheritance of the alleles of two genes are studied together. All the examples you will be given concern genes that are on *different homologous chromosomes*. The inheritance of one pair of alleles is therefore not dependent on the inheritance of the other.

# A2 Human Biology

***Tip*** In problems involving dihybrid inheritance, you must consider one feature at a time, applying exactly the same rules as for monohybrid inheritance.

Chromosomes of each homologous pair are separated during anaphase 1 of meiosis. **Random segregation** of the chromosomes means any combination of alleles (carried on the individual chromosomes from the homologous pairs) is possible.

The diagram below shows a typical dihybrid cross.

| Parental phenotype | Purple flowers tall | White flowers short | Plants from pure-breeding lines are cross-pollinated |
|---|---|---|---|
| Parental genotype | PPTT | pptt | Both are homozygous |
| Parental gametes | PT | pt | Gametes are haploid, so contain only one allele from a pair; only one type of gamete from each parent |
| F1 genotype | PpTt | | All F1 plants are heterozygous for both features and so have purple flowers and are tall (purple flower and tall alleles are dominant). These plants are self-fertilised |
| F1 gametes | PT Pt pT pt    PT Pt pT pt | | All F1 plants can produce four types of gamete; meiosis ensures that only one allele from each pair is present in a gamete (two sets are shown to represent male and female gametes) |

F2 genotypes

**Male**

| | Gametes | PT | Pt | pT | pt |
|---|---|---|---|---|---|
| **Female** | PT | PPTT | PPTt | PpTT | PpTt |
| | Pt | PPTt | PPtt | PpTt | Pptt |
| | pT | PpTT | PpTt | ppTT | ppTt |
| | pt | PpTt | Pptt | ppTt | pptt |

This is the standard way of showing all the possible combinations of alleles in the F2

**9** of the combinations contain both dominant alleles and so will be tall with purple flowers

**3** of the combinations contain only the dominant allele for flower colour and so will be short with purple flowers

**3** of the combinations contain only the dominant allele for height and so will be tall but have white flowers

**1** of the combinations contains only recessive alleles and so will be dwarf with white flowers

F2 phenotypes  9 tall purple : 3 tall white : 3 dwarf purple : 1 dwarf white

***Tip*** A cross between individuals heterozygous for both features produces a phenotype ratio of 9:3:3:1. A cross between an individual heterozygous for both features and one showing both recessive features produces a phenotype ratio of 1:1:1:1.

**Links** Mendelian inheritance might be linked to DNA structure (molecular genetics), protein synthesis, mutations and natural selection. It might also be linked to physiological processes, because reactions are governed by enzymes — coded for by genes.

## Comparing results of crosses with predictions — the $\chi^2$ test

For a simple dominant–recessive monohybrid cross between two heterozygous tall pea plants, we would expect a ratio of 3:1. Is a ratio of 72:28 close enough for us to assume that it really is 3:1 and that the difference is due to chance? To test this by $\chi^2$, construct a table like the one below.

**Tip** To calculate the expected values for a genetic cross, divide the total number of offspring according to the predicted ratio.

| Feature | Observed (O) | Expected (E) | (O – E) | (O – E)² | (O – E)²/E |
|---|---|---|---|---|---|
| Tall | 72 | 75 | –3 | 9 | 0.12 |
| Dwarf | 28 | 25 | 3 | 9 | 0.36 |

$$\chi^2 = \Sigma \frac{(O - E)^2}{E} = 0.48$$

The value of $\chi^2$ is 0.48 with one degree of freedom. Degrees of freedom are calculated by the formula, degrees of freedom = $n - 1$, where $n$ is the number of possible different outcomes. In this case there are two possible outcomes — tall or dwarf — so there is one degree of freedom.

**Tip** In a dihybrid cross between individuals heterozygous for both features, there would be four expected outcomes, so there are three degrees of freedom.

The $\chi^2$ value is now checked in a $\chi^2$ probability table.

| Degrees of freedom | Probability, p | | | | | | | | | | |
|---|---|---|---|---|---|---|---|---|---|---|---|
| | 0.99 | 0.98 | 0.95 | 0.90 | 0.80 | 0.50 | 0.20 | 0.10 | 0.05 | 0.02 | 0.01 | 0.001 |
| 1 | 0.000 | 0.001 | 0.004 | 0.016 | 0.064 | 0.455 | 1.64 | 2.71 | 3.84 | 5.41 | 6.64 | 10.83 |
| 2 | 0.020 | 0.040 | 0.103 | 0.211 | 0.446 | 1.386 | 3.22 | 4.61 | 5.99 | 7.82 | 9.21 | 13.82 |
| 3 | 0.115 | 0.185 | 0.352 | 0.584 | 1.005 | 2.366 | 4.64 | 6.25 | 7.82 | 9.84 | 11.35 | 16.27 |
| 4 | 0.297 | 0.429 | 0.711 | 1.064 | 1.649 | 3.357 | 5.99 | 7.78 | 9.49 | 11.67 | 13.28 | 18.47 |
| 5 | 0.554 | 0.752 | 1.145 | 1.610 | 2.343 | 4.351 | 7.29 | 9.24 | 11.07 | 13.39 | 15.09 | 20.52 |
| 6 | 0.872 | 1.134 | 1.635 | 2.204 | 3.070 | 5.35 | 8.56 | 10.65 | 12.59 | 15.03 | 16.81 | 22.46 |

At one degree of freedom, the value of 0.48 falls between the values for $p = 0.5$ and $p = 0.2$. In biology, for results to be statistically *different*, the value would need to *exceed* that for $p = 0.05$. This $\chi^2$ does not, so these differences can be put down to chance.

**Tip** Perfect agreement between O and E would give a $\chi^2$ value of 0. Small values tend to indicate good agreement; larger ones tend to indicate poorer agreement.

## Variation
### Types of variation
**Continuous variation** produces a whole range of outcomes. Features that show continuous variation are often:
- controlled by several *genes* (**polygenic** inheritance)
- easily measured (e.g. height)

**Discontinuous variation** produces distinct categories. Features showing discontinuous variation are often:
- controlled by alleles of a single gene
- not easily measured (e.g. PTC tasting or flower colour)

### Causes of variation
Variation can be caused by genetic factors, the environment or a combination of both.

Genetic variation is caused by:
- crossing over in prophase 1 of meiosis
- independent assortment of chromosomes in anaphase 1 of meiosis
- random fertilisation
- mutation

**Tip** Only mutations create new genes; the other causes of genetic variation create new combinations of existing genes.

Mutations of a single base triplet of DNA are called **point** mutations. The three main kinds are:
- addition — an extra base is added
- deletion — a base is removed
- substitution — the number of bases is unaltered, but one base replaces another

Additions and deletions alter the entire base sequence after the point of the mutation — they are called **frameshift** mutations.

**Links** The base sequence of a gene codes for a specific amino acid sequence in a protein. Additions and deletions alter the amino acid sequence and so a new tertiary structure results. This affects protein function. If the protein is an enzyme, the active site could be non-functional. Substitutions may have no effect on protein structure because the DNA code is degenerate. However, a new amino acid *could* be coded for, which might alter protein structure and function.

## Populations, selection and speciation
### Populations
The variation of a feature showing continuous variation in a population can be defined by the range, the mean ($\bar{x}$) and the standard deviation ($\sigma$).

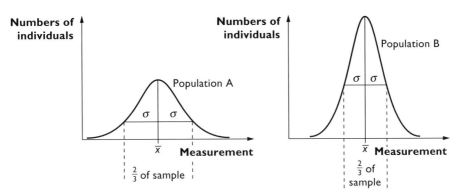

In the above examples, the mean is the same for both populations, but the standard deviation is much less in population B. There is less variation in population B.

*Tip* The standard deviation gives a more useful measure of dispersion about the mean than the range. The range is affected by one or two extreme values; the standard deviation is less affected by these and gives a better idea of 'typical' values.

The total of all the genes of the individuals in a population at a particular time is called the **gene pool**. The frequencies of individual alleles can be calculated using the **Hardy–Weinberg** equation, provided that:
- the population is large
- individual organisms are diploid
- mutation does not occur
- there is no migration
- there is no selection

For a gene with two alleles **A** (dominant) and **a** (recessive), the frequency of **A** is represented by $p$ and the frequency of **a** is represented by $q$.

Equation 1: $p + q = 1.0$
Equation 2: $p^2 + q^2 + 2pq = 1.0$

*Tip* In these equations, $p$ and $q$ must always be expressed as *decimals*. You might have to convert actual numbers of individuals to a frequency. Do *not* calculate a percentage. The value for $q^2$ corresponds to the frequency of *individuals* showing the recessive feature (homozygous for the recessive allele). Taking the square root of this value gives $q$ — the frequency of the recessive allele. Once you know $q$, you can calculate $p$ from equation 1 and then you can calculate everything else.

### Selection

**Natural selection** allows those members of a population that are best suited to a particular condition (the **selection pressure**) to *survive to reproduce*. They pass on the alleles that gave them the advantage to the next generation. Less well-adapted individuals do not survive to reproduce in such large numbers and so fewer of them pass on their alleles to the next generation. The frequency of the advantageous allele increases with each generation.

*Tip* When answering questions on natural selection always:
- identify the selection pressure
- state which type has a selective advantage and why
- explain that these individuals will survive to reproduce and pass on their advantageous alleles
- explain that the frequency of the advantageous genotype and phenotype will increase with each generation

Types of selection include:
- **directional selection**, where one extreme is selected against

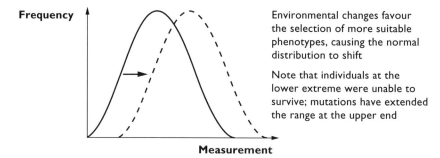

Environmental changes favour the selection of more suitable phenotypes, causing the normal distribution to shift

Note that individuals at the lower extreme were unable to survive; mutations have extended the range at the upper end

- **stabilising selection**, where both extremes are selected against and the mean is favoured

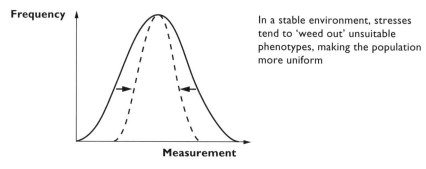

In a stable environment, stresses tend to 'weed out' unsuitable phenotypes, making the population more uniform

- **disruptive selection**, where both extremes are selected for and the mean is selected against

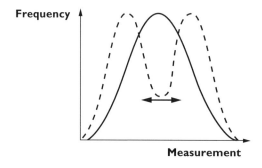

Environmental changes favour the selection of more suitable phenotypes at both extremes of the normal distribution, causing a split

**Links** Natural selection can be linked to questions about Mendelian inheritance. It can also be linked to biochemistry because a mutant allele will produce a different protein, which may give a selective advantage or disadvantage. You might then have to use your knowledge of protein structure or synthesis to explain *why* the protein is different, as well as your understanding of selection to explain any selective advantage.

### Speciation

A species is a group of organisms with a common gene pool that is capable of interbreeding to produce fertile offspring.

*Tip* Do not confuse 'fertile' with 'viable'. Viable just means capable of living, whereas fertile means capable of reproducing.

New species arise from existing ones when a population of a species becomes incapable of breeding with the remaining individuals of that species. This can happen:
- when two populations inhabit different environments and through natural selection acquire different combinations of alleles and features and become incapable of interbreeding (**allopatric speciation**)
- when groups of individuals inhabiting the same environment develop different breeding patterns so that they can no longer interbreed (**sympatric speciation**)

In each case, there is a period of **reproductive isolation**, which allows differences to accumulate between one population and another. Eventually these differences are so great that the populations become different species.

Natural selection provides a mechanism for change, but isolation is the *key stage* in speciation. Isolation allows two populations to change in different ways and eventually to become separate species.

*Tip* Questions about speciation might give data about the ranges of different populations of animals. If the ranges overlap and no intermediate types are found, it suggests that they are distinct species as:
- there is the opportunuty to interbreed (ranges overlap)
- no intermediate types are found (suggests no interbreeding has occurred)

## Classification

Living organisms are classified into five **kingdoms**:
- **Prokaryotae** — unicellular organisms with circular DNA, no membrane-bound organelles, small ribosomes and non-cellulose cell walls.
- **Protoctista** — unicellular and multicellular eukaryotic organisms that are not classified as fungi, animals or plants.
- **Fungi** — unicellular and multicellular, plant-like, non-photosynthetic organisms with eukaryotic cells that have non-cellulose cell walls.
- **Plantae** — multicellular organisms with eukaryotic cells that might contain chlorophyll and have cellulose cell walls.
- **Animalia** — multicellular organisms that develop from a blastocyst and have eukaryotic cells with no cell walls.

A2 Human Biology

The kingdoms are subdivided into ever-smaller groups or **taxa** (hence the term **taxonomy**). For example, the complete classification of the domestic cat is shown below:

| Kingdom | Animalia |
|---|---|
| Phylum | Chordata |
| Class | Mammalia |
| Order | Carnivora |
| Family | Felidae |
| Genus | Felis |
| Species | domestica |

Each organism is **binomial**, i.e. has a name consisting of two parts. The first part is the genus the organism belongs to and the second part is the species. The domestic cat is *Felis domestica*.

## The structure of ecosystems

Ecosystems are self-supporting systems consisting of a **community** of living organisms, made up of **populations** of different species interacting with their **environment**. The environment consists of all the non-living factors or components of the ecosystem. Each species occupies its own **habitat** and **niche** within the environment. A habitat is the physical location in which an organism is found. A niche is the role the organism fulfils within that habitat.

*Tip* Don't forget that the decomposers in an ecosystem are also part of the community.

### Populations

The numbers in a population can be estimated by taking a **random sample** of the population.

For plants or small organisms that do not move around much, the following procedure can be used:
- Place several random quadrats of known area.
- Find the average number of the organism per quadrat.
- Multiply this by the ratio of the area of the site to the area of the quadrat.

*Tip* You cannot *throw* quadrats at random because you must make a choice to throw over your shoulder or in a certain direction. You must use random numbers to define coordinates.

For animals that do move, use the **mark–release–recapture** technique.
- Collect a sample of the animals from the study area, count them ($N_1$), mark them unobtrusively and release them.
- After a suitable period, collect a second sample, count the total number ($N_2$) and the number of marked individuals ($n$).
- The estimate of population size is given by $\frac{N_1 \times N_2}{n}$.

**Links** In a synoptic paper, calculations of population size might be included in questions on ecology and could be related to questions on population genetics and natural selection. Questions using data concerning an unfamiliar organism might include estimation of population size.

### Species diversity

This is a measure that takes into account the **species richness** of an area (the number of different species present) and the relative success of each species.

Simpson's diversity index ($d$) is given by the formula:

$$d = \frac{N(N-1)}{\Sigma n(n-1)}$$

where $N$ = the total number of all organisms in the area and $n$ = the number of organisms of one species in the area.

**Tip** A *high* diversity index suggests a complex, stable ecosystem with many food webs and niches. A *low* diversity index suggests a less complex, less stable ecosystem, probably dominated by just a few species.

**Links** Questions are often linked to succession and to how changes in the environment can affect species diversity.

### Succession

Succession describes the changes that occur as the community and environment change with time. The diagram below shows a succession that begins with an area of bare rock and develops over time into mature woodland.

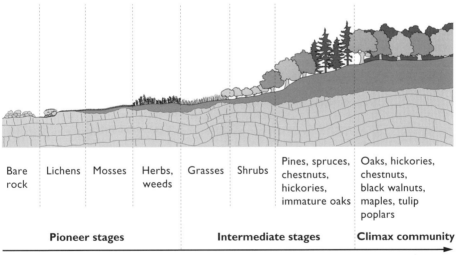

**Tip** Successions like the one shown above can be halted before they reach the climax. For example, grazing prevents shrubs and trees from developing. A fire can destroy an ecosystem without taking it back to its origin. It can subsequently develop along a different succession.

All successions show the following changes with time:
- an increase in species diversity (and species richness)
- an increase in complexity of the community and food webs
- an increased number of ecological niches
- changes in the environment because of changes in the community

**Links** You might have to link changes in a succession to competition for resources. For example, small trees would out-compete grasses for light, water and mineral ions. This could then lead on to photosynthesis and cycling of nutrients.

## Transfer of energy through ecosystems

This involves the transfer of energy *between* organisms as well as between organisms and the environment.
- Photosynthesis by **producers** makes energy-rich organic molecules such as starch.
- Feeding and digestion transfer organic molecules to the **consumers**.
- Death and decay transfer organic molecules to the **decomposers**.
- Respiration releases energy from the organic molecules to 'drive' metabolic processes.

No energy transfer is ever 100% efficient — some energy is always lost as heat. For example, in aerobic respiration only about 50% of the energy stored in a molecule of glucose ends up stored in ATP; the rest is lost as heat in the energy transfer. Also, only a fraction of the light energy striking a chlorophyll molecule becomes stored as ATP in the light-dependent reactions of photosynthesis.

Energy transfer can be illustrated by an **energy flow** diagram:

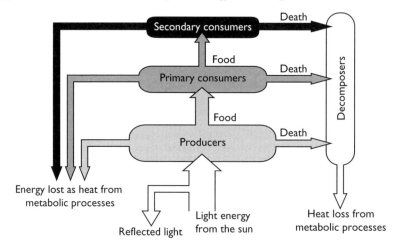

Ecological pyramids based on energy are *always* pyramid shaped. This is because of the energy losses at each trophic level — only about 10% of the energy is passed to the next trophic level. This limits the number of levels in most food chains.

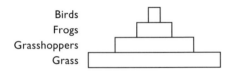 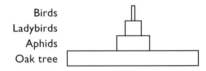

Biomass pyramids are also pyramid shaped. However, the shape of ecological pyramids based on the numbers of organisms in a food chain depends on the size of the organisms involved.

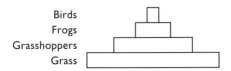 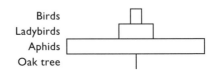

Thousands of grass plants are needed to support the grasshoppers, whereas one oak tree can support millions of aphids.

*Tip* You may have to perform calculations to show how much energy is transferred at various stages. Look at the energy input and outputs at each level. Eventually, they must balance — all the energy entering a level must be passed either to the next level, to the decomposers, or be lost as heat. In any one year, this may not be the case as some of the energy entering a trophic level will be stored in the tissue of those organisms and *not* eaten, decayed or respired.

## Cycling nutrients through ecosystems

### The carbon cycle

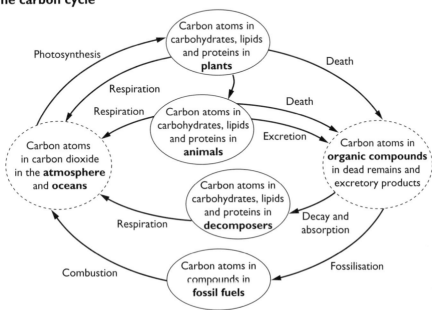

# A2 Human Biology

## The nitrogen cycle

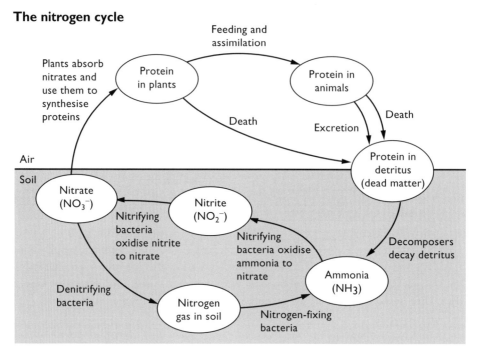

**Tip** Remember that although nutrients are recycled, energy is not. All the energy entering a food chain is eventually lost as heat.

**Links** You might have to give detailed explanations of any of the processes involved in the carbon and nitrogen cycles. The nitrogen cycle can be linked to:
- excretion of nitrogenous waste
- active uptake of nitrate ions by roots
- protein synthesis and structure
- respiration

# Energy transfer within organisms

## The biochemistry of photosynthesis
The process of photosynthesis has two main stages:
- The light-dependent stage uses light energy to generate ATP and reduced NADP. This takes place in the **grana** of a chloroplast. Light energy is transduced into chemical energy in the ATP molecules.
- The light-independent stage uses the ATP and reduced NADP to produce triose phosphate, which can be converted into hexose sugars and then stored as starch. This takes place in the **stroma** of a chloroplast. Energy from ATP molecules is used to synthesise starch molecules.

**Tip** You might have to use Module 1 knowledge to explain why starch is an efficient storage molecule.

# AQA (A) Unit 9a

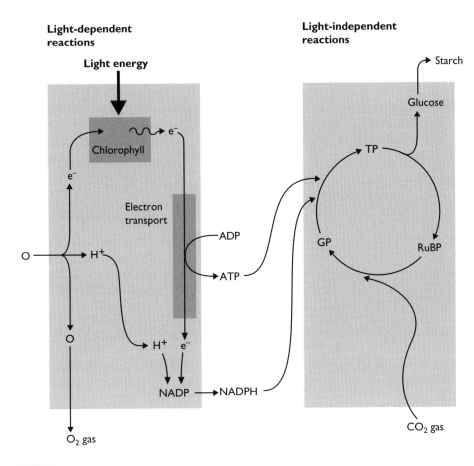

**Links** You might have to relate photosynthesis to:
- crop production and growth (how the factors affecting photosynthesis are controlled to increase productivity)
- the carbon cycle
- the structure of molecules (starch and glucose)
- enzyme activity (all the reactions of photosynthesis are controlled by enzymes)
- genetic control (genes code for proteins that may be enzymes regulating the reactions of photosynthesis)

## The biochemistry of respiration

Respiration releases energy stored in organic molecules such as glucose and uses it to synthesise ATP. **Aerobic** respiration uses oxygen to give a net yield of 38 ATP molecules per molecule of glucose.

Glycolysis takes place in the cytoplasm. The link reaction and Krebs cycle take place in the matrix of the mitochondrion. The molecules of the electron transport chain are located on the inner membrane of the mitochondrion.

A2 Human Biology

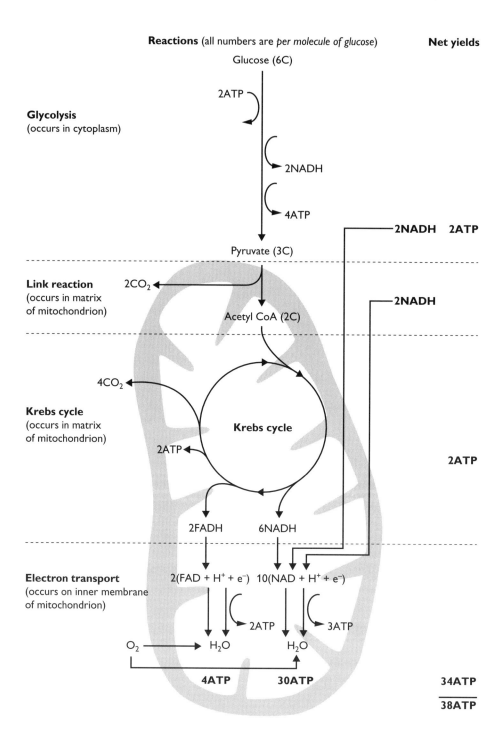

**Anaerobic** respiration gives a net yield of only 2 molecules of ATP per molecule of glucose.

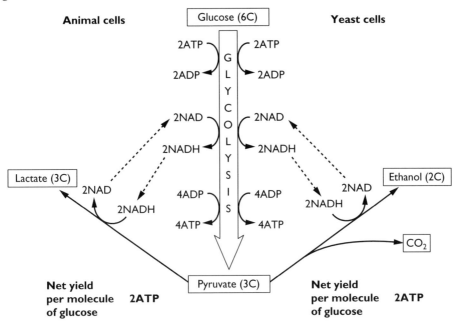

**Tip** When describing the difference in yields of ATP, be careful to use the phrase 'per molecule of glucose'. These are *net* yields because two molecules of ATP are used in each process.

**Links** The biochemistry of respiration can be linked to any energy-requiring process, such as active transport (e.g. the sodium–potassium pump in the membrane of axons, absorption from the gut, soil and nephron), protein synthesis, the Calvin cycle and the first step in glycolysis.

### Respiratory quotients

Respiratory quotient (RQ) is defined by:

$$RQ = \frac{\text{amount of carbon dioxide produced}}{\text{amount of oxygen used}}$$

RQ depends on the respiratory substrate (carbohydrate, lipid or protein) and on any anaerobic respiration taking place.

| Respiratory substrate | RQ |
|---|---|
| Carbohydrate | 1.0 |
| Lipid | 0.7 |
| Protein | 0.9 |

***Tip*** Organisms often respire a mixture of substrates. In this case, the RQ falls between the values for the individual substrates. In plants, if anaerobic respiration is occurring at the same time as aerobic respiration, the RQ is increased because more carbon dioxide is produced but no more oxygen is used. Anaerobic respiration in animals produces no $CO_2$ and so does not influence the RQ in mixed respiration.

### ATP

ATP is produced in respiration and photosynthesis. When hydrolysed by ATPase, it releases energy to drive energy-dependent reactions and processes. It can be resynthesised from ADP and inorganic phosphate using energy from:
- an electron transport chain (in a chloroplast or in a mitochondrion)
- substrate-level phosphorylation reactions (in glycolysis and the Krebs cycle)

**Links** You might be asked why ATP is a suitable molecule for transferring energy within cells.

***Tip*** ATP releases energy in small amounts in a single reaction, and is easily resynthesised in the cell. ATP cannot move out of the cell, but moves around easily within the cell.

## Man's influence on ecosystems: deforestation

Forests are complex ecosystems. Clearing them results in:
- a reduction in species diversity and ecological niches
- a reduction in the rate at which carbon dioxide is removed from the atmosphere, because there are fewer trees photosynthesising (decomposition and/or burning may increase the rate at which $CO_2$ is added to the atmosphere)
- a reduction in the amount of nitrogen returned to the soil (tree trunks removed from the area contain large amounts of nitrogen fixed in proteins and other nitrogenous compounds)
- damage to the structure of soil causing increased leaching of mineral ions

**Links** You might have to compare leaf fall from trees with the application of organic and inorganic fertilisers, or link damage to the soil structure with eutrophication.

Sustainable practices prevent or minimise the changes that take place when forests are cleared. In the example shown on the right:
- each area is felled over a 30-year period, allowing species to migrate to neighbouring areas
- it will be 270 years before that area is felled again, giving time for regeneration
- the central area is never felled, providing a permanent stock of all species
- damage to soil is restricted to one area and is repaired as the area regenerates

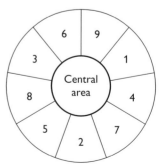

Areas are felled in numerical order

**Links** You might have to interpret diagrams that represent good practice in sustainable felling of forest areas.

# Module 7: The Human Life Span

## Human reproduction — the production and transfer of gametes

### Sperm production (spermatogenesis)

The male sex cells are called **spermatozoa**, often abbreviated to **sperm**. They are produced in **seminiferous tubules** in the **testes** in a process that commences at puberty and continues for the rest of a man's life. There are three main stages in sperm production:

- **multiplication phase** — cells in the epithelium of the seminiferous tubules divide repeatedly by mitosis to form many small cells called **spermatogonia**
- **growth phase** — the spermatogonia grow into **primary spermatocytes**
- **maturation phase** — the primary spermatocytes divide by meiosis (meiosis I) to form **secondary spermatocytes** and then again (meiosis II) to form **spermatids**; the spermatids then develop into mature sperm

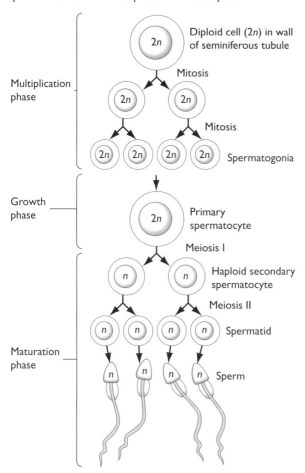

### The production of ova (oogenesis)

The female sex cells — **ova** (singular **ovum**) — are produced by similar processes, but there are some important differences.

Production of ova commences before birth. The multiplication phase and growth phase take place in the immature ovaries of the fetus. The **primary oocytes** (the female gametocytes) begin to divide by meiosis, but then enter a period of 'suspended animation' until puberty. By then they are surrounded by a group of cells and the whole structure is called a **primary follicle**. Then the pituitary hormone **FSH** (follicle-stimulating hormone) stimulates one or more primary follicles to develop each month. Meiosis I resumes and forms one **secondary oocyte** and one **polar body**. Meiosis II commences, but is 'frozen' at metaphase. At this stage, **ovulation** occurs.

Meiosis II is only completed if the nucleus of a sperm enters the secondary oocyte. As the sperm's nucleus enters, meiosis II forms a mature **ovum** and a second polar body. Fertilisation then takes place as the sperm nucleus and ovum nucleus fuse.

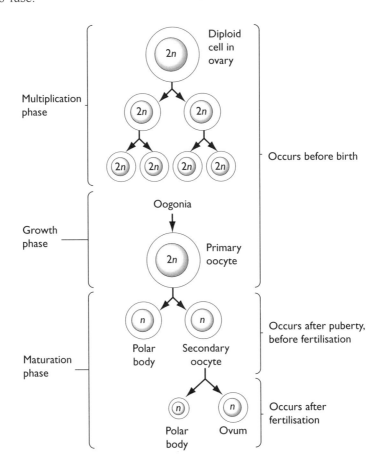

## Differences between spermatogenesis and oogenesis

| Spermatogenesis | Oogenesis |
| --- | --- |
| Continuous process | Periodic process — one primary oocyte develops in each menstrual cycle |
| Commences at puberty | Commences before birth |
| No interruptions | Process halted as meiosis I begins (before birth) and again at metaphase of meiosis II (just before ovulation) |
| Each primary spermatocyte produces four mature sperm | Only one mature ovum and three polar bodies produced |
| Billions of sperm produced in a lifetime | At most 500 secondary oocytes produced in a lifetime |

**Links** A commonly asked question is to present a diagram of spermatogenesis or oogenesis and to ask the candidate to identify the type of cell division occurring at certain stages, linking this to a reduction in chromosome number and genetic variation in the gametes. You may also have to link oogenesis to the hormonal control of the menstrual cycle.

### Fertilisation

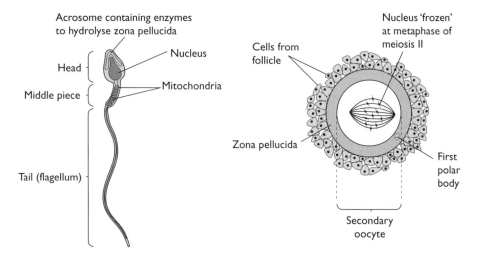

Attracted by a substance secreted by the secondary oocyte, sperm swim towards it and physically push their way through the remaining follicle cells (the **corona radiata**). One sperm will bind to a receptor on the surface of the **zona pellucida**. This initiates the following:
- The **acrosome reaction** takes place; hydrolytic enzymes are released from the acrosome of the sperm and digest a pathway through the glycoprotein zona pellucida.

- The plasma membrane of the head of the sperm fuses with the plasma membrane of the secondary oocyte.
- The plasma membrane of the oocyte draws in the head of the sperm.
- The secondary oocyte completes meiosis II to become an ovum; the second polar body is ejected.
- The nuclei of sperm and ovum fuse to form a **zygote**.
- Cortical granules just beneath the plasma membrane of the ovum secrete enzymes that inactivate the sperm receptors in the zona pellucida; no other sperm can now enter.

**Links** You may have to link the events of fertilisation to:
- hydrolysis (e.g. explain how glycoproteins are hydrolysed)
- protein structure
- the importance of meiosis and fertilisation in a life cycle

### Implantation and early development

The zygote divides repeatedly by mitosis to form first a **morula** (a solid ball of cells) and then a **blastocyst** (a hollow ball of cells).

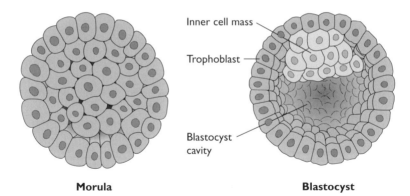

Morula                                   Blastocyst

The inner cell mass of the blastocyst will develop into the **embryo** and the **amnion** (the membrane that will surround the embryo). The outer layer of cells will become the **chorion** (the membrane that forms part of the placenta). The blastocyst normally implants in the uterus lining several days after fertilisation and the chorion then begins to form the placenta.

The placenta is an efficient exchange surface because:
- the chorionic villi create a large surface area
- the chorion is very thin, reducing the diffusion distance between maternal and fetal circulations
- the fetal and maternal circulations maintain concentration differences across the chorion
- fetal haemoglobin has a higher affinity for oxygen than maternal haemoglobin (see page 60) and so can bind with oxygen in conditions under which maternal haemoglobin releases it

**Links** You may be asked to explain the efficiency of the placenta in terms of Fick's law. You may also have to link transfer across the chorion to any of the membrane transport processes.

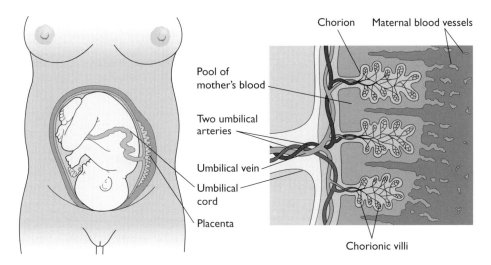

### Placental hormones
The placenta secretes **human chorionic gonadotrophin (hCG)**. This hormone passes into the maternal circulatory system and prevents regression of the **corpus luteum**. The corpus luteum continues to secrete:
- **progesterone**, which maintains the uterine lining and prevents menstruation
- **oestrogen**, which inhibits the secretion of **FSH** by the pituitary gland; this prevents any more follicles from developing

As pregnancy progresses, the placenta secretes less and less hCG and begins to secrete oestrogen and progesterone. The corpus luteum regresses approximately 3 months into the pregnancy.

### Changes in maternal physiology during pregnancy
As the embryo develops and the placenta and uterus enlarge, increasing amounts of oxygen and nutrients are required to meet the demand. Consequently, the following changes take place:
- cardiac output increases
- blood volume increases
- the heart develops more cardiac muscle and the chambers increase in size

As a result of the increased activity, body temperature increases by about 1°C. The increased plasma volume means that water retention can occur. The plasma is more dilute and so the water potential is less negative. Less water is returned osmotically to the plasma from the tissue fluid.

**Links** You may have to link this material to:
- osmosis and water potential, particularly in the context of tissue fluid formation
- cardiac output and its control by the autonomic nervous system

## A2 Human Biology

### Birth and lactation

Just before birth:
- secretion of oestrogen and progesterone by the placenta drops sharply
- the pituitary gland begins to secrete oxytocin

| Event | Consequence |
| --- | --- |
| Decrease in secretion of progesterone | Removes inhibition of contractions of smooth muscle in uterus wall; uterine lining is no longer 'maintained' |
| Decrease in secretion of oestrogen | Removes inhibition of secretion of prolactin |
| Secretion of oxytocin | Actively stimulates contractions of uterus; positive feedback ensures that these become stronger and stronger |
| Secretion of prolactin | Stimulates alveoli in breasts to produce and secrete milk |

After birth, secretion of prolactin and oxytocin continues. Prolactin continues to stimulate alveoli in the breasts to secrete milk, which accumulates in the alveoli. Oxytocin causes this milk to be released into the ducts leading to the nipple, from where the baby can obtain the milk by suckling. Suckling initiates a positive feedback system, which ensures the continued production of milk until the baby is satisfied.

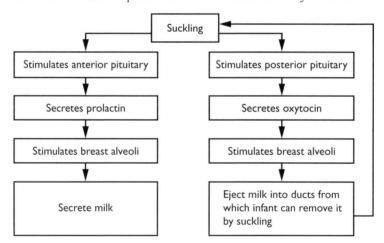

**Links** You may have to link the control of birth and lactation to:
- the tertiary structure of proteins (in the context of receptors for hormones in plasma membranes)
- the mode of action of hormones

### Growth, puberty and ageing

Growth is a permanent increase in the amount of living tissue of an organism. For humans, this means an increase in the number of cells. Growth can be measured in a number of ways.

| Measure | Advantages | Disadvantages |
|---|---|---|
| Standing height | • Convenient<br>• Feet positioned at a fixed point — the ground<br>• Unaffected by metabolism | • Spine is compressed due to the effects of gravity |
| Supine length (lying on back) | • Unaffected by gravity<br>• Unaffected by metabolism | • Inconvenient<br>• Neither feet nor head positioned at a fixed point |
| Body mass | • Convenient<br>• Unaffected by gravity | • Affected by intake of food and water, excretion and egestion |

**Absolute growth** is a measure of the total amount of living tissue year on year, for example height or body mass. **Growth rate** is a measure of the *increase* in living tissue in that year.

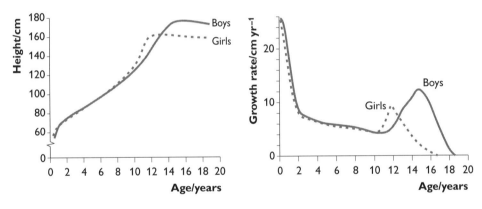

**Relative growth** expresses growth over a year as a fraction or percentage of the amount of tissue at the start of the year. For example, a person who is 150 cm tall grows 7.5 cm in one year:

$$\frac{7.5}{150} \times 100 = 5\%$$

Growth can be measured in **longitudinal studies** (measuring the absolute growth of a group of individuals of the same age over a number of years) or in **cross-sectional studies** (measuring the absolute growth of a number of groups of individuals of different ages at one point in time).

| Type of study | Advantage | Disadvantage |
|---|---|---|
| Longitudinal | • There are no participant variables — the same individuals are used throughout | • Time consuming<br>• Some individuals may drop out, reducing the reliability of the study |
| Cross-sectional | • Data can be collected quickly and conveniently | • There are participant variables — different groups may not have had the same mean height at the starting age |

Tissues and organs grow at different rates:
- The head and brain reach full size by the age of eight, and further development increases the number of nerve connections, allowing completion of increasingly complex tasks.
- Lymphoid tissue develops rapidly in childhood and adolescence. This is linked to combating diseases and immunity; many disease-causing organisms are encountered for the first time during this period of growth.
- Reproductive organs reach full size only after puberty.

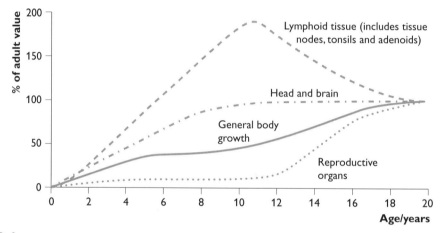

## Puberty

The onset of puberty is marked by an increase in the secretion of **gonadotrophin releasing hormones** by the hypothalamus. These stimulate the pituitary gland to increase its secretion of FSH and lutenising hormone (LH) in both sexes. As a result of these secretions:
- the ovaries increase their secretion of oestrogen and progesterone
- the testes increase their secretion of testosterone

The combined effects of the pituitary hormones and those from the reproductive organs cause:
- the resumption of oogenesis in the ovaries
- the commencement of spermatogenesis in the testes

In addition, the hormones from the reproductive organs stimulate the development of the **secondary sex characteristics**.

| Secondary sex characteristics in males | Secondary sex characteristics in females |
|---|---|
| Widening of the shoulders | Development of the breasts and widening of hips |
| Enlargement of larynx, causing deepening of the voice | Some larynx enlargement and deepening of the voice |
| Growth of facial, body and pubic hair | Growth of pubic hair and underarm hair |
| Enlargement of the sex organs | Onset of menstruation |

The reproductive hormones also stimulate the pituitary gland to increase its production and secretion of **growth hormone**, which causes the **adolescent growth spurt** by increasing bone length, muscle mass and the rate of metabolism.

The **thyroid gland** increases its secretion of hormones, which raises the metabolic rate of the body.

### Ageing

As the young adult ages, all physiological processes decline in efficiency. The rate of decline varies between different organs and systems and between individuals.

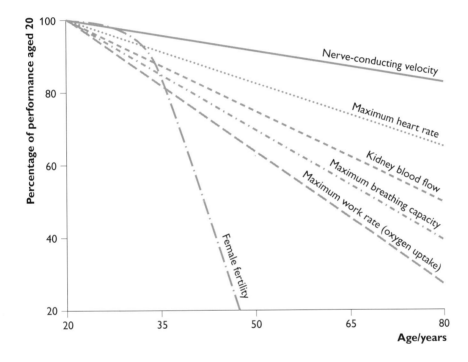

The sharp decline in female fertility at around the age of 35–45 years is called the **menopause**. Follicles in the ovaries become less and less sensitive to FSH and LH. As a result:

- oocytes do not develop and ovulation ceases
- the production of oestrogen and progesterone decreases with the result that levels of FSH and LH remain high (their secretion is no longer inhibited by oestrogen and progesterone)
- consequently, menstruation ceases

**Links** You may have to link material on growth, puberty and ageing to:
- mitosis and meiosis
- cardiac output
- generation and propagation of action potentials and factors affecting the speed of conduction

- pulmonary ventilation
- mode of action of hormones
- tertiary structure of proteins (in the context of receptors for hormones in plasma membranes)
- mean, mode, median, standard deviation and range of groups of individuals used in longitudinal and cross-sectional studies of growth

*Tip* Questions on growth frequently involve calculations. Make sure that you understand the various measures of growth and how to calculate them.

## The transport of respiratory gases in mammals

Most oxygen is transported bound to haemoglobin in red blood cells.

Carbon dioxide is transported:
- in physical solution in the plasma (5%)
- combined with haemoglobin (10%)
- as hydrogencarbonate ions ($HCO_3^-$) in the plasma (85%)

### The haemoglobin dissociation curve

This graph shows the percentage of haemoglobin saturated with oxygen (associated with oxygen to form **oxyhaemoglobin**) at different **partial pressures** of oxygen.

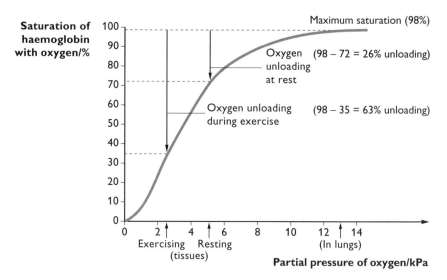

You might have to calculate actual volumes of oxygen being unloaded.

*Tip* To do this, use the following steps:
- Calculate the % haemoglobin unloading (*a*), as shown above.
- If not given, calculate from the data the total volume of oxygen carried (*b*) when haemoglobin is at maximum percentage saturation (*c*).

- The volume of oxygen being unloaded is $a \times \frac{b}{c}$
- Don't forget units!

### Release of oxygen by haemoglobin

The association/dissociation of haemoglobin is a reversible reaction.

$$HbO_8 \rightleftharpoons Hb + 4O_2$$

If the partial pressure of oxygen *or* concentration of free haemoglobin decreases, more oxyhaemoglobin will dissociate. Actively respiring tissues cause both of these effects because the cells:

- use oxygen quickly, reducing the partial pressure of oxygen
- produce carbon dioxide and lactate, leading to the formation of hydrogen ions, lowering the pH; the hydrogen ions bind with haemoglobin (haemoglobin acts as a buffer, preventing the pH from falling too far), reducing the concentration of free haemoglobin

The two effects are independent of each other; lowering pH will cause more haemoglobin to dissociate irrespective of the partial pressure of oxygen.

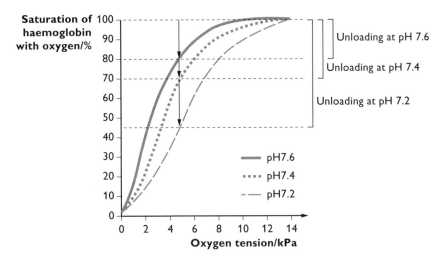

Notice that at any oxygen tension (partial pressure), more haemoglobin dissociates at the lower pH levels. This is the **Bohr effect**.

### Fetal haemoglobin

Fetal haemoglobin has a quaternary structure that varies slightly from that of adult haemoglobin; two of the four polypeptide chains are different, altering its properties. It has a higher affinity for oxygen than adult haemoglobin, meaning that it can associate with oxygen at low partial pressures, when adult oxyhaemoglobin would dissociate.

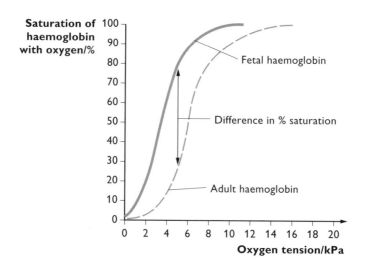

Adult oxyhaemoglobin dissociates at the placenta, unloading oxygen, which diffuses across the chorionic membrane and associates with fetal haemoglobin. This association maintains a low partial pressure of free oxygen in the fetal plasma at the placenta, maintaining the concentration gradient that allows oxygen to continually diffuse across the chorion.

**Links** You may have to link the transport of respiratory gases to:
- the structure of proteins, with particular reference to haemoglobin
- the inheritance and effects of abnormal haemoglobin (e.g. thalassaemia, sickle-cell anaemia)
- mutations (producing abnormal haemoglobin)
- diffusion and Fick's law
- aerobic respiration

## Digestion and absorption in humans

### Digestion in humans

| Gut region | Secretion | Enzyme(s) | Substrate(s) | Digestion product(s) |
|---|---|---|---|---|
| Buccal cavity (mouth) | Saliva | Amylase | Starch | Maltose |
| Stomach | Gastric juice | Pepsin | Proteins | Short-chain polypeptides |
| Lumen of duodenum and ileum | Pancreatic juice | Amylase | Starch | Maltose |
| | | Lipase | Triglycerides | Fatty acids and glycerol |
| | | Trypsin | Protein | Short-chain polypeptides |
| | | Exopeptidases | Short-chain polypeptides | Dipeptides |
| Ileum wall | | Maltase | Maltose | α-Glucose |
| | | Dipeptidase | Dipeptides | Amino acids |

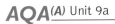

**Links** All reactions of digestion involve **hydrolysis**. You might have to link a digestive process to knowledge about how bonds are broken by hydrolysis (Module 1). Digestion might also be linked to factors affecting enzyme action, particularly pH, since this varies through the gut.

### Absorption of the products of digestion
Absorption from the lumen of the ileum into surrounding epithelial cells occurs by:
- diffusion (fatty acids and glycerol)
- facilitated diffusion (glucose and amino acids)

Concentration gradients for these processes are maintained by:
- exocytosis of **chylomicrons** (containing the recombined fatty acids and glycerol) into **lacteals** (lymph vessels)
- active transport of amino acids and glucose from the epithelial cells into the space between these cells and the capillaries; from here they diffuse into the capillaries

**Links** You might have to relate the method of absorption of a particular molecule to its structure and to the fluid mosaic model of plasma membranes.

The ileum is adapted for efficient absorption by having:
- a large surface area (due to its length, circular folds in the internal wall, villi and microvilli)
- thin epithelial cells, which create a short diffusion/transport distance
- a good blood supply in which the circulation maintains a concentration gradient between the epithelial cells and the plasma

## Dietary requirements
### Nutrients in our diet

| Nutrient | Function(s) in body |
| --- | --- |
| Carbohydrates | • Starch is hydrolysed to glucose which is oxidised in respiration to produce ATP |
| Lipids | • Respiratory substrate<br>• Phospholipids in plasma membranes<br>• Stored in adipose tissue to give thermal insulation<br>• Schwann cells give electrochemical insulation to axons |
| Proteins | • Manufacture of enzymes<br>• Manufacture of actin and myosin for muscles<br>• Transport proteins, channel proteins in plasma membranes<br>• Manufacture of some hormones (e.g. insulin, glucagon) |
| Vitamins | • Vitamins have no overall function, but individual vitamin types are important for very specific benefits in the body (e.g. Vitamin D is important in absorption of calcium ions from the gut and a deficiency in childhood causes rickets) |
| Minerals | • Iron is essential in the formation of haemoglobin and cytochromes (components of the electron transport chain in the inner membrane of mitochondria)<br>• Calcium is essential for: formation of bones and teeth; transmission of nerve impulse across a synapse; contraction of skeletal muscle |

The proportional requirements of the various nutrients vary during our lives as lifestyle and body composition change.

### Estimating the energy content of foods

The energy content of food is estimated by **calorimetry**. A sample of food is burned and the energy released used to heat a known mass of water. The energy content of the food per gram is given by the equation:

$$\text{energy released (joules per gram)} = \frac{\text{final temperature} - \text{initial temperature} \times 4.2}{\text{mass of water (g)}}$$

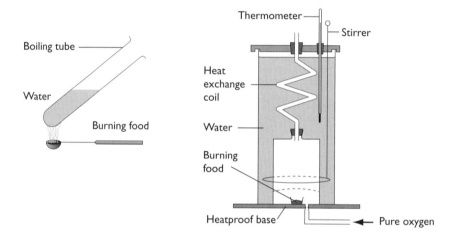

The second method gives a more reliable estimate because:
- less heat from the burning food escapes into the air
- the stirrer ensures a more even distribution of heat
- the food is burned in pure oxygen, ensuring a more efficient combustion and releasing more energy

### Testing foods for the main nutrients

| Nutrient | Reagent used | Procedure | Result |
|---|---|---|---|
| Reducing sugar | Benedict's solution | Heat sample of food with Benedict's solution | Yellow/red precipitate |
| Non-reducing sugar | Benedict's solution | • Heat sample of food with Benedict's solution<br>• Hydrolyse by boiling with HCl<br>• Neutralise by adding NaHCO$_3$<br>• Re-test with Benedict's solution | • No change in colour with initial test<br><br><br>• Yellow/red precipitate |
| Protein | Biuret solution | Mix sample with Biuret solution and allow to stand 1–2 minutes | Lilac/mauve/purple coloration forms |
| Lipid | Ethanol and water (the emulsion test) | • Shake sample with ethanol (lipids are soluble in ethanol)<br>• Pour ethanol into water | Milky/white/cloudy emulsion forms |

## Special diets

| Type of diet | Benefits | Problems | Notes |
|---|---|---|---|
| Weight loss | Reduces risk of obesity, heart disease and diabetes | Diet may not supply all nutrients unless planned carefully | Need to change eating habits to prevent putting weight back on |
| Vegetarian | • Reduced intake of saturated fat<br>• Can feed more people as less energy loss along food chain | • Many non-animal foods are low in protein<br>• Protein may not contain all essential amino acids, so must eat a range of foods | Transamination can produce any non-essential amino acids not supplied in the diet |
| Glycogen loading (athletes) | Maximises glycogen stores in muscle prior to an event; these can be hydrolysed to glucose and then respired to release energy | — | Athletes often have large muscle blocks and so also need larger amounts of protein in their diet |
| Menstruating female | Diet should include extra iron to allow production of red blood cells lost in menstruation | IUDs (coils) can increase menstruation and so the need for extra iron is even greater | Oral contraceptives decrease menstruation, reducing the need for extra iron |
| Pregnancy and lactation | Diet should include extra protein, carbohydrate, calcium and iron | • High-energy snacks should be avoided, as should alcohol<br>• Foods rich in iron and folic acid should be eaten in larger quantities | • Protein for growth of fetus, placenta and milk formation<br>• Carbohydrate for growth and increased work load (carrying extra weight)<br>• Calcium for growth of teeth and bones of fetus<br>• Iron for formation of fetal red blood cells and extra maternal red blood cells |

**Links** You may have to link material on diet to:
- the structure of molecules of carbohydrates, lipids and proteins
- condensation and hydrolysis
- the nature of growth and its measurement
- aerobic and anaerobic respiration
- the cause and incidence of heart disease

## Receptors in humans

Receptors are **energy transducers**. Some kind of energy (e.g. light/heat) leads to a **generator potential** being produced, which *might* initiate an **action potential** in a

**neurone** (nerve cell) with which the receptor forms a **synapse**. The passage of an action potential along the axon of the neurone is called a **nerve impulse**.

A larger input of energy to a receptor results in a larger generator potential, but action potentials are *all the same magnitude*. A larger generator potential results in *more* (not bigger) action potentials.

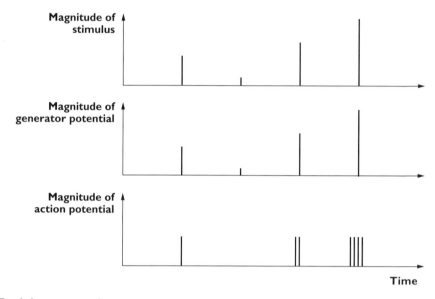

### Pacinian corpuscles

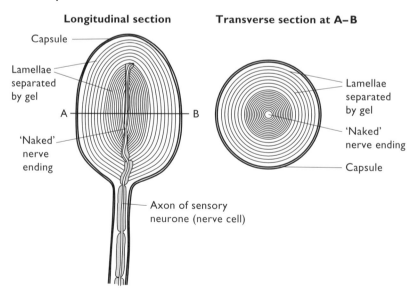

Pacinian corpuscles transduce mechanical force (pressure) into generator potentials. Pressure activates pressure-sensitive sodium ion channels in the membrane of the nerve ending. These open and sodium ions flood through them, altering the charge on the membrane. This change is the generator potential. Increased pressure opens more ion channels, resulting in a greater generator potential.

### Rods and cones in the eye
Rods and cones transduce light energy into generator potentials.

**Rods** contain a light-sensitive pigment called **rhodopsin**. When light strikes rhodopsin, it splits into two substances (opsin and retinal), causing a change in the membrane potential — the generator potential. ATP is used to resynthesise the rhodopsin.

*Tip* This is an important use of ATP. You may have to relate the high numbers of mitochondria in the rods to the resynthesis of rhodopsin.

There are three types of **cone** containing pigments sensitive to different wavelengths of light, broadly corresponding to blue, green and red. Stimulating different types or combinations of types of cone results in the perception of different colours.

**Sensitivity** is the ability to detect low-intensity light. **Acuity** is the ability to distinguish between points close together. Rods and cones have different sensitivity and acuity because of the way in which they are linked to bipolar cells in the retina. Each cone is linked to one bipolar cell; *several* rods are linked to one bipolar cell (**retinal convergence**). In low-intensity light, sub-threshold stimulation from several rods can combine to exceed the threshold and initiate an action potential in a neurone, giving high sensitivity. However, the acuity is poor because the brain is unable to distinguish between the points of the image received by each rod. Cones have low sensitivity but high acuity.

### How the eye produces an image on the retina
Rays of light from a point on an object are brought to focus at a single point on the retina. This happens because parts of the eye **refract** (bend) the rays of light, as shown below.

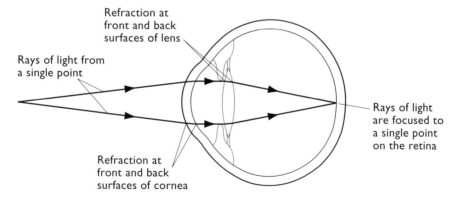

The ability of the eye to change focus from near to distant objects and back again is called **accommodation**.

| Structure | Near accommodation | Distant accommodation |
|---|---|---|
| Cornea | Fixed curvature and so the same refraction of light in near and distant accommodation | |
| Ciliary muscles | Contract[1] | Relax[2] |
| Suspensory ligaments | Slacken[1] | Become taut[2] |
| Lens | More convex[1] | Less convex[2] |

[1] These changes increase the angle (amount) of refraction to bring sharply diverging rays of light to focus on the retina

[2] These changes decrease the angle of refraction to bring less sharply diverging rays of light to focus on the retina

## Transmission of information through the nervous system

### Transmission along neurones

Nerve impulses occur when action potentials are propagated along the axons of neurones. Action potentials are generated when the normal **resting potential** of the membrane of an axon is disturbed sufficiently. (Resting potential is an unfortunate term since the axon membrane is *not* at rest — ions are being moved across the membrane in both directions, by several processes.)

The resting potential of an axon membrane is maintained so that the inside of the axon membrane is 70 mV more negative than the outside. It is maintained in the following way:

- Large, negatively charged particles (anions) are present inside the axon.
- The sodium–potassium pump pumps out three sodium ions for every two potassium ions it pumps in. Both sodium and potassium ions carry a single positive charge.
- Sodium ions *diffuse in more slowly* than potassium ions diffuse out.

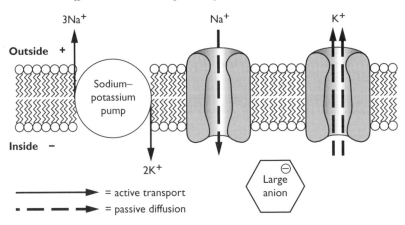

*Tip* When you are describing the movement of ions across the axon membrane, make it clear that you are referring to ions. Sodium is a metal and pieces of metal do not move across the membrane. Also, make clear what sort of movement you are referring to.

**Neurotransmitters** secreted by other neurones or by receptors alter the permeability of the axon membrane to sodium ions, resulting in more sodium ions diffusing *in* through the ion channels. This raises the potential of the inside of the axon membrane. If this reaches the **threshold** value of −55 mV, then thousands of **voltage-sensitive** sodium ion channels open and sodium ions flood in. This raises the potential inside the axon membrane from −55 mV to +50 mV. This change in membrane polarity is called **depolarisation** and the potential of +50 mV on the inside of the axon membrane is the **action potential**.

Before this part of the axon membrane can generate another action potential, it must first be restored to the resting potential. This is **repolarisation**. The sodium ion channels close and the potassium ion channels open. Potassium ions diffuse out, making the inside of the axon more negative again. The time between one action potential and a return to the resting potential is called the **refractory period**. No action potentials can be generated during this period. Therefore, the duration of the refractory period determines how many nerve impulses can pass per second.

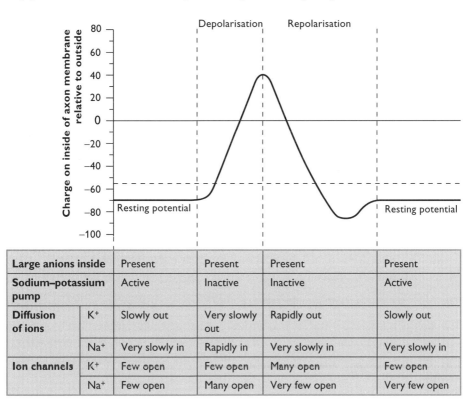

| | | | | | |
|---|---|---|---|---|---|
| **Large anions inside** | | Present | Present | Present | Present |
| **Sodium–potassium pump** | | Active | Inactive | Inactive | Active |
| **Diffusion of ions** | K+ | Slowly out | Very slowly out | Rapidly out | Slowly out |
| | Na+ | Very slowly in | Rapidly in | Very slowly in | Very slowly in |
| **Ion channels** | K+ | Few open | Few open | Many open | Few open |
| | Na+ | Few open | Many open | Very few open | Very few open |

An action potential produces **local currents**, which cause changes in the next part of the axon membrane, resulting in depolarisation. This generates an action potential in this part of the membrane which is then propagated along the axon.

*Tip* You should be able to relate changes in membrane potential to events in the membrane.

The speed of conduction of nerve impulses depends on:
- the diameter of the neurone — there is less resistance to transmission in larger diameter neurones and impulses travel faster
- myelination — myelinated axons are capable of saltatory conduction, in which action potentials are only generated at the nodes of Ranvier. The impulse jumps from node to node, and so conduction is quicker. Myelinated axons expend less energy than non-myelinated ones, as they only reset the original sodium and potassium ion balance using the sodium–potassium pump at the nodes of Ranvier
- temperature — all physiological processes are controlled by enzymes and are therefore temperature-dependent

### Transmission across synapses

Dendrites of two neurones — the **presynaptic neurone** and the **postsynaptic neurone** — make up a synapse. The gap between the two is the **synaptic cleft**.

*Transmission across an excitatory synapse*

An action potential arriving at the membrane of the presynaptic neurone of an excitatory synapse causes the following events:
- Calcium ions enter the neurone, causing vesicles containing neurotransmitter to move towards the presynaptic membrane.

*Tip* Do not confuse this with the entry of sodium ions into an axon.

- The vesicles release the neurotransmitter into the synaptic cleft and it diffuses across the cleft to the postsynaptic membrane.
- The neurotransmitter binds to specific protein receptor sites on the postsynaptic membrane, changing the membrane potential.
- If sufficient neurotransmitter has been released, it raises the membrane potential above the threshold value and initiates an action potential in the axon of the postsynaptic neurone.
- The neurotransmitter is hydrolysed and the components are secreted into the synaptic cleft. They diffuse back across the presynaptic membrane.
- In the presynaptic neurone, ATP is used to resynthesise the neurotransmitter.

*Tip* Note that this is another important use of ATP.

Transmission across a synapse is *unidirectional* because neurotransmitter is secreted only by the presynaptic membrane and receptors are present only in the postsynaptic membrane.

**Links** Questions on synapses might require you to link binding to specific receptors to your knowledge of protein structure. Do not confuse binding sites with active sites. You might also have to explain the process of hydrolysis.

### Inhibitory synapses

The neurotransmitter released at these synapses makes the membrane potential more negative than −70 mV. This makes it less likely that the threshold level will be reached and an action potential generated in the postsynaptic neurone following subsequent excitatory stimulations.

### Summation at synapses

Sometimes, action potentials in the presynaptic neurones produce only a **sub-threshold** stimulus in the postsynaptic neurone and so no action potential is initiated.

If several neurones form synapses with just one other neurone, there may be a mixture of excitatory and inhibitory synapses. If several of these synapses release neurotransmitter at the same time (**spatial summation**), or if one neurone releases several lots of neurotransmitter in very rapid succession (**temporal summation**), their effects add together or **summate**. Two or more sub-threshold stimuli at excitatory synapses can combine to produce a stimulus great enough to reach the threshold. However, just one inhibitory synapse can have a significant effect.

The table below shows possible effects of summation at synapses on the generation of an action potential in the postsynaptic neurone. Neurones A, B and C all form synapses with neurone D.

**Links** You could be given a similar table without the information on which synapse is excitatory and which is inhibitory and asked to complete the table.

| Neurone | Action potential | | | | |
|---|---|---|---|---|---|
| A — excitatory synapse | ✗ | ✓ | ✗ | ✓ | ✓ |
| B — excitatory synapse | ✗ | ✗ | ✓ | ✓ | ✓ |
| C — inhibitory synapse | ✗ | ✗ | ✗ | ✗ | ✓ |
| D — postsynaptic neurone | ✗ | ✗ | ✗ | ✓ | ✗ |

*Tip* Stimulation from neurones A and B together is needed to produce an action potential in D. They must be sub-threshold and excitatory. Stimulation by neurone C cancels these effects, so this must be an inhibitory synapse.

## The autonomic nervous system

The autonomic nervous system has two divisions:
- the **sympathetic** nervous system, which prepares the body for action
- the **parasympathetic** nervous system, which restores the resting conditions after action

Neurones of the sympathetic division release **noradrenaline** at synapses. Those of the parasympathetic division release **acetylcholine**.

The two divisions have **antagonistic** effects.

# A2 Human Biology

| Feature | Sympathetic | Parasympathetic |
|---|---|---|
| Cardiac output | Increases | Decreases |
| Diameter of arterioles leading to skin | Decreases | Increases |
| Skeletal muscle | Contracts | Relaxes |

You need to know all the effects in the table above, but you only have to be able to *explain* the control of cardiac output. You need to be able to explain the effects of both divisions on:
- heart rate, as a result of increased stimulation of the SA node
- stroke volume

**Links** You might need to include details about the AV node, Purkyne tissue and bundles of His from Module 1.

## Reflex actions

Reflex actions result from nerve impulses passing along specific sequences of neurones called **reflex arcs**. **Somatic reflexes** control movement by skeletal muscles (e.g. knee-jerk, withdrawal and blinking). **Autonomic reflexes** control functions such as heart and breathing rates.

The diagram shows the somatic reflex arc controlling a **withdrawal reflex**.

*Tip* A reflex *arc* is a series of structures that allows a reflex *action* to be carried out.

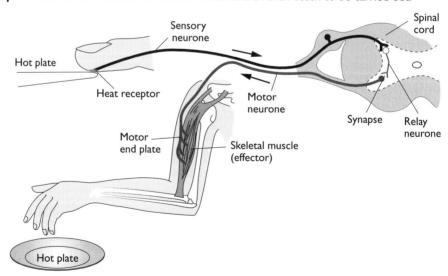

Reflex actions:
- are often protective
- always produce the same response to the same stimulus
- are automatic, requiring no conscious thought or decision

*Tip* Any action with a choice to be made between possible outcomes *cannot* be a reflex action.

**Links** Questions on reflex actions can link to transmission of nerve impulses and transmission across synapses, as well as to topics such as active transport and the specific structure of protein receptors.

## Control by hormones

Hormones are produced by **endocrine glands**. They are secreted into the bloodstream and travel to **target organs** to produce their effects. They are able to target specific organs because of precisely shaped protein receptor molecules in the plasma membranes of target organ cells.

*Tip* A protein receptor has a specific tertiary structure that will only allow a hormone with a complementary shape to bind to it. The tertiary structure is a consequence of the primary structure, which is genetically determined.

Responses controlled by hormones are:
- slower to begin than those controlled by nerves (extra time is needed for circulation and for sufficient hormone to bind to receptors)
- longer-lasting than responses controlled by nerves (hormone production does not stop immediately and hormone molecules remain bound to receptors, so the effects last for some time)

Hormones are either **steroids** (e.g. sex hormones) or **non-steroids** (e.g. insulin and glucagon). They influence cells in different ways:
- Steroid hormones have small, lipid-soluble molecules which can enter cells. They bind with receptors on the nuclear envelope and activate specific genes in the nucleus.
- Non-steroid hormones bind with receptors on the plasma membrane. This stimulates the production of cyclic AMP (**cAMP**) from ATP, which then acts as a **second messenger** and activates enzymes in the cell.

*Tip* Note that this is another use of ATP.

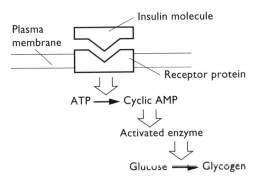

**Links** You might have to use your knowledge of the structure of plasma membranes to explain why steroid hormones are able to enter cells and non-steroid hormones cannot.

# A2 Human Biology

## Control of digestive secretions by nerves and hormones

The diagram shows how both nerves and hormones are involved in the control of gastric secretions.

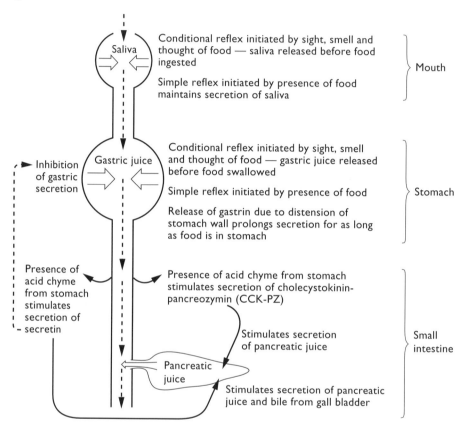

You might be asked to explain the advantage of having more than one method of controlling a secretion.

*Tip* The conditioned reflex secretion of saliva ensures it is in the mouth before food enters, so digestion can begin immediately. Nervous control of gastric secretion ensures quick initial production of gastric juice. Hormonal control ensures that secretion continues for as long as food is in the stomach.

## The structure and function of skeletal muscle

### Contraction of skeletal muscle

Each individual muscle (e.g. the biceps) is an organ, made from the following tissues:
- skeletal muscle — to provide the ability to contract
- blood — to supply the oxygen and glucose needed for aerobic respiration
- nerves — to control the rate and intensity of contraction

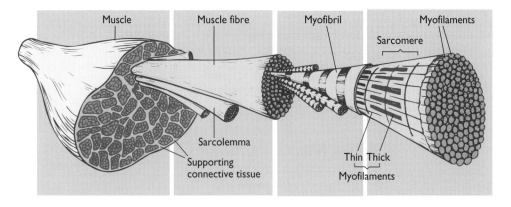

To contract, the filaments of actin and myosin slide relative to one another. Contraction takes place when:
- a nerve impulse causes the release of calcium ions around the actin molecules
- the calcium ions cause **tropomyosin** to move and expose the binding sites on the actin molecules
- ATP bound to heads of myosin molecules is hydrolysed and the energy released is transferred to the myosin heads, which bind with the exposed binding sites on the actin molecule
- the myosin heads move (using the energy from ATP), moving the actin filaments
- the myosin heads detach, another molecule of ATP binds to each and the cycle repeats (as long as nervous stimulation continues)

**Differences in appearance between relaxed and contracted skeletal muscle**

The diagrams below represent muscle in the relaxed and then contracted states.

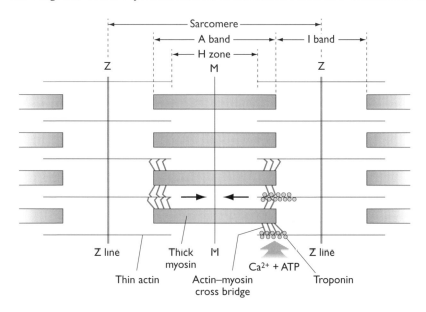

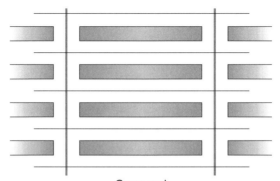

Contracted
(note that there is now no H zone and a much reduced I band)

| Feature | Relaxed | Contracted |
|---|---|---|
| Sarcomere | Longer | Shorter |
| H zone | Present | Absent/reduced |
| I band | Large | Absent/reduced |

### The release of energy during exercise

ATP is made available for muscle contraction:
- directly, from aerobic and anaerobic respiration
- by the **creatine phosphate** pathway from aerobic and anaerobic respiration

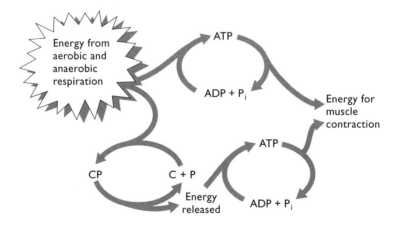

During exercise, energy is released in the following ways:
- Initially, the small store of ATP is used.
- Creatine phosphate is hydrolysed and the energy released used to synthesise more ATP.
- As the exercise progresses, ATP is made available from aerobic and anaerobic respiration.

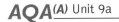

Unit 9a

The muscles of endurance athletes contain a large proportion of slow-twitch fibres, whereas those of athletes involved in 'explosive' events (sprints, weight-lifting) contain a large proportion of fast-twitch fibres.

| Feature | Slow-twitch fibres | Fast-twitch fibres |
| --- | --- | --- |
| Number of mitochondria | Many | Fewer |
| Concentration of Krebs cycle enzymes | High | Low |
| Amount of myoglobin in fibres | High | Lower |
| Respiration | Mainly aerobic | Mainly anaerobic |
| Resistance to fatigue | High | Low |
| Contraction rate | Slow | Fast |

These features allow slow-twitch fibres to continue contracting for long periods as they respire mainly aerobically and produce little lactate, the principal cause of fatigue in muscles.

**Links** You may have to link muscle structure and function to:
- cells, tissues and organs (related to the structure of a skeletal muscle)
- changes in the pattern of circulation of blood during exercise
- changes in cardiac output and pulmonary ventilation during exercise
- aerobic and anaerobic respiration
- the transport of respiratory gases, including the Bohr effect

## Homeostasis

Homeostasis involves maintaining a 'steady state'. It keeps the tissue fluid surrounding cells at a constant concentration, temperature and pH. This allows enzymes to function efficiently and the metabolic reactions they control to take place quickly.

In mammals, the liver is an important homeostatic organ because:
- it generates heat, which is important in maintaining a constant, high body temperature
- it is involved in the control of plasma glucose concentration
- it deaminates surplus amino acids, producing the nitrogenous excretory product urea

Many homeostatic mechanisms operate by **negative feedback**, in which deviation from the norm initiates responses that restore the norm.

### Controlling body temperature

Humans are **endotherms**, like all mammals. They regulate their body temperature by physiological means, despite changing environmental temperature.

The flowchart shows the main physiological processes involved in temperature regulation in humans. Changes in the blood and/or skin temperature initiate responses that restore the temperature to within the set range. Therefore, negative feedback is involved.

# A2 Human Biology

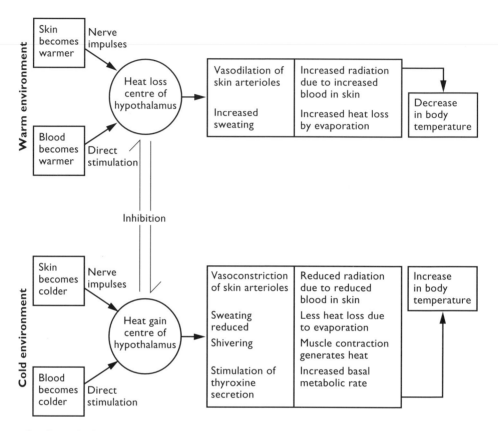

The hypothalamus can also send impulses to areas of the brain that initiate behavioural responses, such as putting on or taking off clothes, or having a cool or warm drink.

**Links** Questions on temperature regulation might involve linking the structure of arterioles to how they are dilated or constricted to redistribute blood. You might have to give specific examples of negative feedback from data supplied.

*Tip* Be precise when describing temperature control. The hypothalamus detects changes in *blood* temperature — 'body temperature' is too general a term.

### Controlling plasma glucose concentration

Plasma glucose concentration tends to rise following a meal because carbohydrate is digested and any glucose formed is absorbed. It tends to fall between meals because respiring cells absorb glucose from the plasma.

α-**cells** in the **islets of Langerhans** in the pancreas respond to a decrease in plasma glucose concentration and secrete **glucagon** to restore the balance. β-**cells** respond to an increase in plasma glucose concentration and secrete **insulin**.

The flowchart below shows how these two hormones interact to maintain the plasma glucose concentration within narrow limits.

# AQA (A) Unit 9a

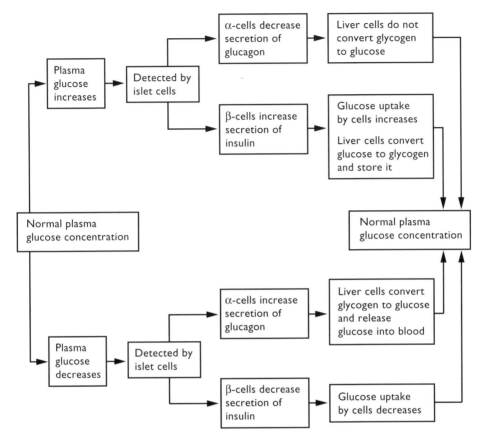

**Links** Questions on control of plasma glucose concentration might be linked to:
- digestion and absorption of carbohydrates
- respiration
- mode of action of insulin or glucagon
- protein receptors for insulin or glucagon, related to tertiary structure of proteins and complementary shapes. *Be careful not to mention active sites.*

You might have to give specific examples of negative feedback from data supplied.

People who cannot produce sufficient insulin suffer from **diabetes mellitus**. The raised plasma glucose concentration that results can cause:
- excretion of glucose in the urine
- production of large volumes of urine, resulting in thirst
- fatigue, as body cells are less able to absorb glucose

Simply adjusting the amount of carbohydrate in the diet can control some forms of diabetes. More severe forms require injections of insulin.

**Links** You might have to use your knowledge of osmosis to explain why excretion of glucose in the urine results in large volumes of water being lost.

A2 Human Biology

# Thinking synoptically

Synopsis means bringing together ideas to give a general overview of a subject. In the introduction to the content guidance section (page 8) there is a synopsis of the content of each module. The main topics are brought together in a single, short paragraph. In synoptic questions, you are required to bring together ideas about a particular topic, for instance ATP, from different areas of the specification.

There are synoptic questions in the Unit 7 and Unit 9a tests.

## Unit 7

Synoptic elements of most questions in the Unit 7 test are generally quite short and straightforward. For example, in a question about muscle contraction, you might be asked to:

**Explain how the pattern of circulation is changed to make extra oxygen available to muscles.**

This is asking you to relate your knowledge on circulation from Module 1 to the process of muscle contraction.

You could include these ideas:
- arterioles leading to skeletal muscle dilate, increasing the blood flow to muscles
- arterioles leading to the skin and gut constrict, decreasing the blood flow to these areas
- cardiac output is increased

In a question on digestion, you might be asked to:

**Complete the diagram to show how a triglyceride molecule is broken down into fatty acids and glycerol.**

You would need to include precise detail from Module 1 of how hydrolysis of a triglyceride occurs.

$$\begin{array}{c} H & O \\ | & \| \\ H-C-O-C-R^1 \\ | & O \\ | & \| \\ H-C-O-C-R^2 + 3H_2O \\ | & O \\ | & \| \\ H-C-O-C-R^3 \\ | \\ H \end{array} \xrightarrow{hydrolysis} \begin{array}{cc} H & O \\ | & \| \\ H-C-OH \quad HO-C-R^1 \\ | & O \\ | & \| \\ H-C-OH \quad HO-C-R^2 \\ | & O \\ | & \| \\ H-C-OH \quad HO-C-R^3 \\ | \\ H \end{array}$$

Ester bond

***Tip*** You *must* show the role of the water molecules.

The last question in the Unit 7 test requires you to write at some length. There will probably be three sections, each worth 5–6 marks. One or more of these sections might be fully or partly synoptic. For example, in a question about the nervous system, the following might form one section:

**Explain how the sympathetic and parasympathetic divisions of the autonomic nervous system interact to influence cardiac output.**

This requires you to explain what effect each division has on the SA node and how the effect is brought about. You would then need to explain how the activity of the SA node affects cardiac output. All this is based on knowledge and understanding of the control of heart rate from Module 1.

You would need to include these ideas:
- cardiac output = heart rate × stroke volume
- The sympathetic and parasympathetic nervous systems act antagonistically.
- The sympathetic division secretes noradrenaline at the SA node, causing it to increase heart rate and stroke volume.
- The parasympathetic division secretes acetylcholine, causing the SA node to decrease heart rate and stroke volume.

## Unit 9a

### Question 1

This is a synoptic structured question. Most sections are worth 2–4 marks and are fairly straightforward. For example, part of a question could be about temperature control in unfamiliar animals:

**The diagram shows a polar bear, which must survive sub-zero temperatures in the Arctic, and a sun bear, which lives near the equator in Asia.**

Polar bear                    Sun bear

**(i) Explain how the relative size of each animal is an adaptation to its environment.**

You would need to include these ideas:
- The polar bear is the larger animal.
- The larger animal has smaller SA/V ratio.
- Therefore, the polar bear will lose less heat through its surface.

**Tip** You could make the reverse argument for the sun bear.

A2 Human Biology

**(ii) Suggest how natural selection could have resulted in the increased size of polar bears.**

You would need to include these ideas:
- Mutation(s) gave rise to increased size.
- Increased size had survival value. Therefore, more of the larger animals survived to reproduce.
- They passed on their advantageous genes.
- The next generation had more large bears.
- This was repeated in subsequent generations.

These questions use the theme of different size bears to bring together very different areas of the specification — the relationship between size and SA/V from Module 1 and natural selection from Module 5.

## Question 2

This will be a comprehension question using a stimulus passage on probably unfamiliar material. It will have every fifth line numbered so that you can refer easily to specific points in the text.

**The following is part of a comprehension passage on tropical rainforest.**

*Tropical rainforest*

The canopy of the rainforest is made up of the upper branches and leaves of the tallest trees. In many places, little light passes through this canopy. There are few plants growing at ground level. Where there are breaks in the canopy, there is a profusion of smaller trees and shrubs. This adds to the species richness and species
5   diversity of the rainforest.

The soil underneath these massive trees is quite shallow; it is stabilised by the roots of the trees. It is also relatively nutrient-poor. Most of the nutrients are 'locked up' in the trees. Leaf-fall, followed by decay, releases enough nitrates and other mineral ions for the following year's growth. Without this rapid recycling,
10   the soil would not be able to support the rainforest.

***Tip*** The first few sentences are all concerned with trees and light and you should be clued in to photosynthesis. At the end of the paragraph the emphasis shifts to species diversity. The second paragraph is concerned with recycling mineral ions, particularly nitrogen. You should be thinking about the nitrogen cycle. It also mentions the role of roots in stabilising the soil, so there is a potential question about what might happen if the trees were felled.

**(a) (i) Why are there few plants growing at ground level beneath the canopy of the rainforest?** *(lines 2–3)* (2 marks)

***Tip*** You must 'lift' the relevant information (there is little light) from line 2 and then use it to explain the lack of plants (little photosynthesis is possible).

**(ii) Why do the gaps in the canopy result in greater species richness and diversity?** *(lines 3–5)* (2 marks)

*Tip* You must relate the information to your knowledge of species diversity. More ecological niches are available.

**(iii) Suggest what is meant by the phrase 'most of the nutrients are "locked up" in the trees'.** *(lines 7–8)* (2 marks)

*Tip* You are being asked to *suggest*, so you are not expected to *know* the answer. You are expected to make an intelligent suggestion from the information supplied. In this case, nitrates are only available from leaf-fall and decay. Most nitrogen is 'locked up' in proteins, DNA and other organic nitrogen-containing compounds in the living cells of the trees.

There is a full comprehension question, together with candidates' answers and examiner's comments, in the Question and Answer section.

## Question 3

There will be a choice between two essay titles. This is where you are most likely to have to pull together ideas from all areas of the specification, so you need to have a suitable strategy. You could think of the modules as the different areas, or you could think in terms of these topics:
- cells and biochemistry
- physiology
- ecology
- genetics, selection and evolution
- disease

Your essay should include material from most modules or most topics.

Two essay titles that have already been used are:
- The different ways in which organisms use ATP.
- How the structure of cells is related to their function.

Let's see how you could pull together different ideas for each of these essays.

### The different ways in which organisms use ATP

You should start with a *short* introduction about how ATP is formed, just to set the scene.

*Tip* You should *not* write *in detail* about respiration — it's not relevant.

Then explain *why* ATP is used as an immediate source of energy and give examples of ATP use, such as:
- in regenerating rhodopsin
- in DNA replication
- in the synthesis of proteins and other macromolecules
- in muscle contraction
- in glycolysis

A2 Human Biology

- in the light-independent reactions of photosynthesis
- producing cyclic AMP (cAMP)
- in active transport. Describe some examples, such as:
  - the sodium–potassium ion pump in nerve cells
  - uptake of mineral ions from the soil
  - uptake of glucose or amino acids from the ileum
  - uptake of glucose or amino acids from the first convoluted tubule

***Tip*** Be careful not to spend too much time describing several examples of active transport in detail, because you are really only describing the same process in different contexts.

These ideas span all four modules, but only include two of the four areas: cells and biochemistry, physiology, ecology, genetics, selection and evolution. This doesn't matter. We have brought together material from different parts of the specification and so have the basis of a synoptic essay.

### *How the structure of cells is related to their function*
There are many examples of specialised cells in the specification. In this essay, you must relate the structures of several different cells to their functions and say how a particular structure enables a cell to carry out a particular function.

***Tip*** There is no place in this essay for detailed descriptions of individual organelles.

You should include the adaptations of some of the following cells and relate the adaptations to specific functions:
- red blood cells (no nucleus, haemoglobin, biconcave disc)
- mammalian nerve cells (long axon, myelination, dendrites producing synapses)
- cells lining the first convoluted tubule (microvilli and many mitochondria)
- skeletal muscle 'cells'
- rod cells (rhodopsin)
- cone cells (three different light-sensitive pigments)

***Tip*** Be careful here: the connection of rods and cones to nerve cells in the retina is not really part of their *structure*.

- cells in the alveolar epithelium (thin)
- cells lining the ileum (microvilli and many mitochondria)
- palisade cells (tubular shape, many chloroplasts)
- spongy cells (irregular shape)

***Tip*** Close packing of palisade cells and loose packing of spongy cells is not relevant unless you have described the feature (shape) that allows this.

- root hair cells (hair-like extensions, mitochondria)
- xylem (lignified walls)
- guard cells (shape, uneven thickening of cell walls)

***Tip*** Opening and closing of stomata by guard cells is not an acceptable answer, unless you describe the features that make this possible.

There are other types of cell, but these are the ones that you are expected to know most about. You could not possibly include all of them in the time available.

## Planning an essay

The stages you must go through are as follows:
- Read each essay title carefully.
- Read it again carefully.
- Read it again carefully.
- Read it once again just to make sure!

*Tip* It is absolutely essential that you are clear in your mind about what is wanted. For example, 'The transfer of energy through ecosystems' is not a general essay about ecosystems; it is much more precise.

- Start to make a list of the topics that you consider relevant.
- If these lead on to other related ideas, then also note these. However, be sure that they are still related to the title and not merely to one of the topics you have noted. You might produce a 'spider diagram', similar to the one shown below for the ATP essay.

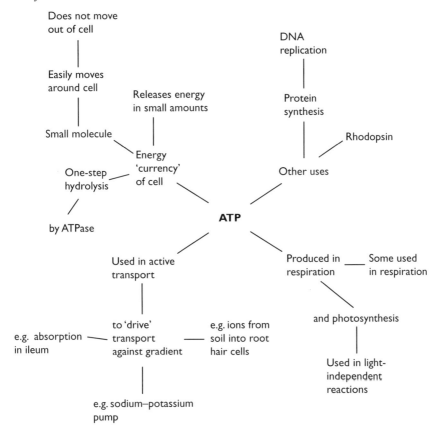

- Now that you have your ideas, decide how much detail you can include about each.
- Arrange your ideas into a logical sequence. Make your plan legible, so that if you run out of time the examiner can look at your plan and read what you were planning. You might get some credit for good intentions!

*Tip* At this stage, check once again that what you are intending to write does relate to the title of the essay.

- Start to write your essay.

## Writing an essay

To understand the advice here, you need to know how your essay will be marked. This is different from marking the answers to a structured question, where the examiner is looking only for the correct biology. Here, you will be credited (or not) for the way you construct the essay *as well as* for the appropriate biology.

*Tip* You are not expected to write a biological thesis or to use the kind of expressive language that you might in an English essay. You must write clearly and logically using scientific terminology. The amount you write will vary according to the size of your writing and the nature of the essay. However, if you do not include too much padding, you ought, with average sized writing, to be able to score full marks in $2\frac{1}{2}$–3 sides of paper.

The marks are for:
- scientific content (0–16 in 2 mark increments — 2, 4, 6, etc.)
- breadth (0–3)
- relevance (0–3)
- quality of language (0–3)

Therefore, the essay is marked out of a total of 25 marks.

To make sure you get the highest possible mark you should:
- try to include as many different aspects of the topic as you can — you will then be awarded 2 or 3 out of 3 marks for breadth
- make sure that *all* the material is relevant to the title — you will then be awarded 3 marks for relevance
- try to write coherently, using scientific terminology wherever you can, so that your essay is easy to follow — you will then be awarded 2 or 3 out of 3 marks for quality of language

The mark you score for scientific content will be determined by how much you know, which is determined by how well you have prepared for the examination. However, bear in mind that:
- the content must be of genuine A-level standard
- it must be relevant — if you write brilliant A-level biology that is not relevant to the essay title, you will receive no credit for it and you will be penalised under the 'relevance' section

**Tip** You do not have to include *every* detail to score 16 marks. The examiners are aware of the pressure you are under in examinations.

## Possible synoptic titles for essays

Writing this list probably puts the kiss of death on any of these titles ever being used in an examination. However, they are genuine synoptic essay titles that were used before this guide was written.

- The structure and function of proteins
- The structure and function of carbohydrates
- The structure and function of lipids
- The transfer of energy within and between organisms
- The uses of ATP
- The ways in which cells are adapted to their functions
- How humans influence the environment
- Maintaining a constant internal environment
- The importance of nitrogen to living organisms
- Carbon in living organisms
- How mutations can result in the production of non-effective enzymes
- The importance of membranes in and around cells

There are many other titles that could be used, but these should give you a flavour of the type of essay that could be set. Try to produce a plan for each. Even if these titles are never set, putting together synoptic essay plans is a valuable exercise.

# Questions & Answers

This section contains questions similar in style to those you can expect to see in your Unit 9a examination. The limited number of questions means that it is impossible to cover all the topics and all the question styles, but they should give you a flavour of what to expect. The responses that are shown are real students' answers to the questions.

The Unit 9a test has 55 marks. There will be:
- a structured question, worth 15 marks
- a comprehension question, worth 15 marks
- an essay question, worth 25 marks

There will be a choice of essay title.

There are several ways of using this section. You could:
- 'hide' the answers to each question and try the question yourself. It needn't be a memory test — use your notes to see if you can actually make all the points you ought to make
- check your answers against the candidates' responses and make an estimate of the likely standard of your response to each question
- check your answers against the examiner's comments to see if you can appreciate where you might have lost marks
- check your answers against the terms used in the question — did you *explain* when you were asked to, or did you merely *describe*?

## Examiner's comments

All candidate responses are followed by examiner's comments. These are preceded by the icon *e* and indicate where credit is due. In the weaker answers, they also point out areas for improvement, specific problems and common errors such as lack of clarity, weak or non-existent development, irrelevance, misinterpretation of the question and mistaken meanings of terms.

# Question 1

## Structured question

Cystic fibrosis is an inherited condition in which epithelial cells, especially those lining the bronchioles, produce increased amounts of mucus that is more viscous than normal.

(a) Cystic fibrosis is determined by a single allele. It is estimated that, in England, one person in 2500 suffers from the condition. The pedigree below shows the pattern of inheritance of cystic fibrosis in one family.

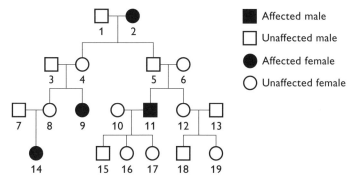

(i) What evidence in the pedigree supports the view that the allele causing cystic fibrosis is recessive? (1 mark)

(ii) What is the probability that another child born to individuals 3 and 4 will be a male with cystic fibrosis? Explain your answer. (2 marks)

(iii) Calculate the frequency of people in England who are heterozygous for this condition. (2 marks)

(b) The normal allele codes for the production of a channel protein (CFTR) in the plasma membranes of epithelial cells. This protein transports chloride ions out of the cells. In cystic fibrosis sufferers, the recessive allele codes for a non-effective protein. The diagram below shows how transport of ions through epithelial cells is altered in cystic fibrosis sufferers.

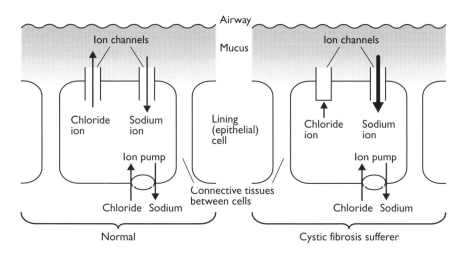

# A2 Human Biology

**question**

(i) Explain how inheriting a different allele results in a non-effective protein being synthesised. (3 marks)

(ii) Suggest why cystic fibrosis sufferers produce mucus that is more viscous than normal. (3 marks)

(c) Attempts are being made to treat other genetic diseases by genetically engineering lymphocytes to produce proteins that will prevent or limit the disease.

(i) Explain how a gene probe could be used to identify the gene. (2 marks)

(ii) Explain how the gene could be removed from the lymphocyte cell. (2 marks)

**Total: 15 marks**

■ ■ ■

*e* (a) Pedigrees are convenient ways of presenting a lot of genetic information. When you read about numbers of people suffering from a genetic condition, you should be preparing mentally for a question involving the Hardy–Weinberg equations.

(i) This is a standard question. There is only one answer — and you should know where to look.

(ii) You must know the genotypes of 3 and 4. Now draw a genetic diagram. Remember that only half the offspring will be male.

(iii) You *must* start with the frequency of the individuals showing the recessive trait, i.e. $q^2$ in the Hardy–Weinberg equations.

(b) The information links alleles to protein structure. You should be thinking about the different levels of protein structure. The diagram shows how a different protein results in different ion and water transport. You should be thinking about water potential gradients and osmosis.

(i) You should be able to relate a different base sequence to different protein structure.

(ii) Viscosity must be linked to water content and water movement by osmosis.

(c) (i) You should be able to recall the action of gene probes from Module 3.

(ii) This requires direct recall from Module 3.

■ ■ ■

## Candidates' answers to Question 1

### Candidate A
**(a) (i)** Parents 1 and 2 have more unaffected than affected children, so the cystic fibrosis allele must be recessive.

### Candidate B
**(a) (i)** 7 and 8 are normal and 14 is affected. Therefore, 7 and 8 must be heterozygous, carrying the dominant allele.

# AQA (A) Unit 9a

> 🅮 Candidate A does not know where to look. The only unequivocal proof of an allele being recessive is a child showing a feature not shown by the parents. Candidate B provides the correct evidence and scores the mark.

**Candidate A**

(a) (ii) 7 and 8 are heterozygous, so there is a 1 in 4 chance of the next child having cystic fibrosis.

**Candidate B**

(a) (ii)

|   | X | Y |
|---|----|----|
| X | XX | XY |
| X | XX | XY |

Male, $p = 0.5$

|   | N | n |
|---|----|----|
| N | NN | Nn |
| n | Nn | nn |

Affected, $p = 0.75$

Overall, $0.5 \times 0.75 = 0.375$

> 🅮 Candidate A has not taken into account the gender of the child. Candidate B shows an understanding of how to work out the probability, but has calculated the probability for an unaffected male, not for a male with cystic fibrosis. Both candidates gain 1 mark for partially correct answers.

**Candidate A**

(a) (iii) $p^2 + q^2 + 2pq = 1$
$p + q = 1$
$p = 1 - q$
1 in 2500 = 0.0004
$p = 0.9996$
$2pq = 2 \times 0.9996 \times 0.0004 = 0.0008$

**Candidate B**

(a) (iii) 1 in 2500 = 0.0004 = $q^2$
$q = 0.02$, so $p = 1 - 0.02 = 0.98$
$2pq = 2 \times 0.98 \times 0.02 = 0.0392 = 3.92\%$

> 🅮 A question on allele frequencies will always give you information that enables you to calculate $q^2$. You can then calculate $q$ and all other values. Candidate A knows the Hardy–Weinberg equations, but not how to apply them, and scores no marks. Candidate B understands how to use the equations and scores 2 marks.

**Candidate A**

(b) (i) This is because a different amino acid is produced.

**Candidate B**

(b) (i) The base sequence after the mutation is altered and now codes for a different sequence of amino acids. Therefore, a different protein is made.

> 🅮 Both candidates appreciate that different amino acids will be produced. Candidate B explains that this is because the base sequence is altered. Candidate A

# A2 Human Biology

scores 1 mark and Candidate B scores 2 marks. Neither candidate explains that a different tertiary structure would make the protein non-effective.

**Candidate A**

(b) (ii) The chloride ions don't leave the cell and so they don't affect the concentration.

**Candidate B**

(b) (ii) In a cystic fibrosis sufferer, the chloride ions don't leave the cell because the protein pump is ineffective. More sodium ions enter the cell, so there are more ions in the cell. This draws water in by osmosis from the mucus in the airways.

> *e* Candidate A only mentions the change in chloride ion movement and scores 1 mark. Candidate B covers the change in movement of both ions, but does not link this to reduced water potential in the cell. However, the candidate does realise that water is withdrawn from the mucus to make it more viscous and that this must involve osmosis. Candidate B scores all 3 marks.

**Candidate A**

(c) (i) A gene probe is radioactive and has a matching base sequence to the gene you are looking for. If you mix a gene probe with some DNA and it stays radioactive, the DNA must have the gene.

**Candidate B**

(c) (i) The gene probe is radioactive. When you add it to a DNA sample it binds, leaving the sample radioactive. This can be checked with a Geiger–Muller tube. If radioactivity is found, the sample must have the gene.

> *e* Both candidates understand about gene probes being radioactive and that this provides a method of detection. However, neither gives a correct explanation of why the probe binds to the gene. Candidate A comes nearest, but does 'matching' mean an *identical* sequence or a *complementary* sequence? It is not clear. Both candidates score 1 mark.

**Candidate A**

(c) (ii) The gene can be cut out with a DNA ligase enzyme to leave sticky ends.

**Candidate B**

(c) (ii) You could use a restriction endonuclease to cut out the gene. This would leave sticky ends on the DNA so that it could be easily inserted into the host DNA.

> *e* Both candidates clearly understand the procedure, but Candidate A has given the wrong enzyme and so only scores 1 mark. Candidate B scores 2 marks.

> *e* Candidate A has made some basic errors that with more thorough preparation would not have been made. There is a general lack of detail and clarity in some of the answers. In contrast, Candidate B uses precise language and gives more detail. Candidate A scores 5 out of a possible 15 marks. Candidate B scores 12.

# Comprehension question

**Read the following passage.**

All the exchange surfaces between blood and cells of mammals involve capillaries. The walls of these tiny vessels are made of a single layer of flattened cells called squamous endothelium. The blood flows very slowly through the capillaries of an organ because of the large collective surface area.

5 At the arterial end of a capillary network, tissue fluid is forced from the capillaries. Protein molecules are not forced out with the tissue fluid and play an important part in the return of water to the plasma at the venous end of the network. Some of the fluid that leaves the capillaries does not return at the venous end. Instead, it drains into lymph vessels and is returned to the blood near the heart.

10 In the glomerular capillaries in the kidney, ultrafiltration forces fluid through a basement membrane into the lumen of the renal capsule. From here, the filtrate passes along the nephron, where many substances are reabsorbed. The variable reabsorption of water in the collecting ducts is used to control the water potential of the blood plasma. Some desert mammals, such as camels, can produce extremely concentrated
15 urine to minimise water loss in times when water intake is low. Camels have all their body fat stored in a hump on their backs. It was once thought that this was an important source of metabolic water. However, calculations show that camels would lose more water vapour by breathing in extra oxygen to oxidise the fat than could be gained by respiring the fat. There must be another biological reason for nearly all the body
20 fat being located in one place.

**(a)** Describe three features that allow capillaries to exchange substances efficiently with tissues. *(lines 2–4)* (1 mark)

**(b) (i)** Explain why proteins are not forced out of the blood at the arterial end of capillary networks. *(lines 5–7)* (1 mark)

**(ii)** Explain the importance of protein molecules in the return of water molecules to the plasma at the venous end of the network. *(lines 6–7)* (2 marks)

**(c)** The diagram below shows the relationship between lymph vessels and blood capillaries.

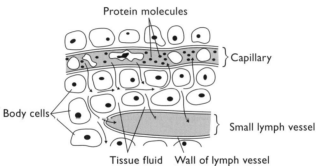

**question**

(i) There is no pump in the lymphatic system. Suggest why lymph moves along the lymphatic vessels. (2 marks)

(ii) In some African countries, a type of parasitic nematode worm blocks the lymphatic vessels in the legs. Suggest why tissues in the lower regions of the legs become swollen. (4 marks)

(d) (i) The formula of starch is $(C_6H_{10}O_5)_n$; the formula of a typical triglyceride lipid is $C_{17}H_{35}COOH$. Use this information to explain why a camel would need extra oxygen to respire lipids aerobically. (*lines 18–19*) (3 marks)

(ii) Suggest a biological benefit to a desert mammal of having all the body fat located in one place in the body. (*line 20*) (2 marks)

**Total: 15 marks**

■ ■ ■

In the first paragraph, all the material is directed at the structure of capillaries and exchange. You should be mentally relating the two as you read. The second paragraph is about the formation of tissue fluid. You should relate this to its composition as you read. The third paragraph concerns the kidney. Towards the end of the paragraph, the emphasis shifts to desert mammals. Clearly, the camel is not specifically included in your specification, but you are being given a number of cues to topics that are. You should be thinking about respiration as well as about ways in which locating the fat in a hump may minimise water loss.

(a) Here, you are being told *exactly* where to look for straightforward information, so there is only 1 mark.

(b) (i) You should be aware that the basis of filtration is molecular size.

(ii) You should be aware that, if no physical force is involved, water moves by osmosis down a water potential gradient.

(c) (i) You do not *know* this, so the information *must* be in the diagram or paragraph.

(ii) You need to make connections. Look at the diagram — if the movement *along* the lymph vessels is blocked, this must limit movement *into* them.

(d) This is straightforward recall of knowledge from Module 7.

(e) (i) Specifying aerobic respiration should cue you to consider the use of *oxygen*.

(ii) Body fat in a mammal should make you think of insulation. Removing it from most of the body removes that insulation.

■ ■ ■

## Candidates' answers to Question 2

### Candidate A

(a) They are made from squamous epithelium and blood flows slowly through them. They have a large surface area.

### Candidate B

(a) They are thin-walled, have a large surface area and only let the blood flow slowly. This gives more time for exchange.

# AQA (A) Unit 9a

> 🅔 Both candidates are awarded the mark, although the answer from Candidate B is more accurate. Candidate A does not quite make it clear that thinness is a feature. However, as the passage describes squamous cells as flattened and the candidate clearly identifies both other features, the mark is awarded.

**Candidate A**

(b) (i) Proteins are not forced out because the pressure is too high.

**Candidate B**

(b) (i) Protein molecules are not forced out because they are too large.

> 🅔 Candidate A does not understand the concept and does not score. Candidate B gains the mark.

**Candidate A**

(b) (ii) Protein molecules are soluble in water and so attract the water to them by osmosis.

**Candidate B**

(b) (ii) The plasma proteins create a water potential gradient between the tissues and the plasma (higher in the tissues). Water moves down the gradient.

> 🅔 Candidate A does not seem to understand the concept, but has mentioned that water moves by osmosis. Candidate B understands the role of plasma proteins, but has forgotten to mention that water moves by osmosis. Both candidates gain 1 mark.

**Candidate A**

(c) (i) It is pushed along when more lymph enters the lymphatic vessels from the blood.

**Candidate B**

(c) (i) As more fluid enters from the extracellular tissue fluid, the increased pressure forces the lymph along.

> 🅔 Both candidates understand the idea of increased pressure. However, Candidate A's answer is not strictly accurate as what enters the lymphatic vessels is *not* lymph and it does not enter directly from the blood. Candidate A scores 1 mark and Candidate B scores 2.

**Candidate A**

(c) (ii) The worms will block the lymph vessels so the lymph can't flow along them and there will be swelling in the tissues.

**Candidate B**

(c) (ii) As worms block the lymph vessels, they prevent or slow down the flow of lymph away from the tissues. This prevents more fluid from entering the lymph and so there is swelling.

> 🅔 Candidate A only explains reduction in lymph flow and scores 1 mark. Candidate B also makes this point but goes on to say that this prevents further fluid entering. Candidate B scores 2 marks. Neither mentions that *increased pressure* in the

# question 2

lymph prevents further fluid entry or that some fluid cannot return to the blood. Neither really *explains* the swelling.

**Candidate A**

**(d) (i)** You have to breathe in more oxygen to respire triglycerides because they are bigger molecules and so more oxygen is needed. The oxygen is the terminal hydrogen acceptor.

**Candidate B**

**(d) (i)** In respiration, carbon is oxidised to carbon dioxide and hydrogen is oxidised to water. The carbohydrate already contains some oxygen, but the triglyceride only contains a little and so the camel would need to breathe in more.

> *e* Candidate A is generously awarded 1 mark for giving a use of oxygen. However, no comparison is made between the oxygen content of the two molecules. Candidate B explains how oxygen is used and why more is needed to oxidise triglycerides. Candidate B scores all 3 marks.

**Candidate A**

**(d) (ii)** If all the body fat is in one place, then the other parts of the body have no fat and so the body will not get as hot. It will be able to cool down.

**Candidate B**

**(d) (ii)** Body fat acts as insulation. Without this insulation, some parts of the body will be able to radiate heat outwards much more easily and so cool the body down. The camel will have an increased chance of surviving in a hot desert.

> *e* Candidate A only gives part of the story. Why will the body be able to cool down? There is no mention of the *insulating* properties of fat or that without it, *radiation* of heat can occur more easily. Candidate A has not given any of the necessary detail and scores no marks. Candidate B scores both marks.

> *e* **Candidate A appears to understand many of the ideas involved in this question, but loses marks by using imprecise language or by failing to supply detail. This is particularly true in the later sections with higher mark allocations. By contrast, Candidate B not only understands the ideas, but is well prepared and gives most of the relevant detail. Candidate A scores 5 out of 15 marks. Candidate B scores 12.**

# Essay question

Write an essay on *one* of the following topics:
(a) **The structure and functions of proteins** (25 marks)
(b) **The transfer of energy within and between organisms** (25 marks)
You should select and use information from different parts of the specification. Credit will be given not only for the biological content, but also for the selection and use of relevant information, and for the organisation and presentation of the essay.

■ ■ ■

> *e* The instructions are there to remind you of the marking criteria. Don't forget the advice on how to plan and write an essay (pages 83–85).

■ ■ ■

## Candidates' answers to Question 3

### Candidate A
**(a)** I am going to write about how proteins are made up and the ways in which living things use them. Proteins are very important molecules and have a variety of uses in living things. They are used for growth and for repair as well as for making enzymes. First I shall write about what proteins are made from.

> *e* A brief introduction is a sensible idea, but this is not very good. The first sentence is just saying 'I am going to write about what you asked me to write about'. There is no need for this.

All proteins contain nitrogen. Some contain carbon, hydrogen, oxygen and phosphorus as well (CHONPS). The building blocks of proteins are amino acids. These are joined together in long chains to form polypeptides, which are the proteins.

> *e* The first few lines here are really only GCSE standard and the wording is somewhat confused.

The sequence of the amino acids in a protein molecule is determined by the sequence of bases on the DNA in the nucleus. Each group of three bases (called a triplet) codes for one amino acid. The code is degenerating and some amino acids have more than one code. A sequence of bases in the DNA molecule makes a corresponding sequence of bases called mRNA. This carries the code out of the nucleus to the ribosomes and the ribosomes read the code. Transfer RNA molecules bring the correct amino acids and they are joined together.

> *e* This is not relevant to this essay. The title only asks for the structure and functions of proteins. It is good biology (though what is a *degenerating* code?) but is costing the candidate marks, because he/she will be penalised under the 'relevance' section of the mark scheme.

# A2 Human Biology

## question

When two amino acids are joined together, they are joined by a peptide bond. This keeps on happening until all the amino acids have been joined and the chain is complete. This is called the primary structure of the protein.

> 🅔 There is no mention that peptide bonds are formed by condensation.

When the primary structure has been formed, it curls into a spiral structure called an α-helix. This is held in shape by bonds called hydrogen bonds between some of the amino acids. This is called the secondary structure of the protein. It can then fold itself into an even more complicated shape called the tertiary structure. This is a 3-D shape and each protein has its own tertiary structure.

> 🅔 There is no mention of the disulphide bonds and ionic bonds that also hold the tertiary structure in shape.

Because it has its own shape, the protein is individual. There is not another one like it. Proteins carry out specific functions because of their shapes. Enzymes have an active site, which is a specific shape. Because of this shape, it can only react with one substance to form an enzyme–substrate complex. This is why enzymes are specific. If enzymes are heated too much, their 3-D shape is altered because some of the bonds are broken and the active site alters. Because the active site isn't the same shape any more, it can't fit with its substrate and it is denatured. This also happens with pH.

> 🅔 This is mostly correct, but it is not only enzymes that are denatured by heat — all proteins are. What does the last sentence mean?

Some of the proteins in the plasma membrane have active sites that substances can bind to. This allows hormones like insulin to recognise their target cells. Insulin can only bind to cells with the right receptors on their surface because they have the correct shape active site to fit the shape of the insulin molecule. Adrenaline and glucagon work the same way.

> 🅔 This is a wrong use of the term 'active sites'. Only enzymes have active sites. These plasma membrane proteins have **binding sites**.

Some proteins on the plasma membrane are antigens. These are only found in one person. Everybody else's antigens are different. This means that your immune system can recognise antigens that shouldn't be there because they are a different shape from the ones that should be there. Antigens sometimes have carbohydrates attached to them or lipids. Antibodies are also proteins and each antibody has its own active site that binds with one specific antigen and destroys it.

> 🅔 Antibodies have **binding sites**, *not* active sites.

Proteins are also used to produce urea in the liver. The proteins are digested to amino acids in the small intestine and then absorbed. They travel to the liver in the hepatic portal vein and are then absorbed by the hepatocytes (liver cells). The amino acids are deaminated to form ammonia and a keto acid. The ammonia is reacted with carbon dioxide in the ornithine cycle to form urea. The urea travels to the kidney and is excreted.

*e* This is irrelevant and part of it is wrong. Proteins are not *used* to make urea. Surplus amino acids are deaminated to form urea. The rest is good biology, but because it is irrelevant it is costing the candidate marks.

Sometimes proteins are used for respiration, if all the other reserves have been used up. Proteins have an RQ of 0.7, so you can tell if an organism is respiring protein or carbohydrate.

*e* There is no need at all to quote RQ. If you must, then get it right! Proteins have an RQ of 0.9.

Proteins are used to make enzymes and for growth and repair and so they are an important part of our diet. We need to eat about 60 g of protein every day.

In conclusion, proteins are made from amino acids and have special shapes that make them have certain functions. Haemoglobin is a protein with iron in it that can transport oxygen in human blood.

*e* This is mostly a repeat of earlier information, which cannot score marks. Right at the very end, another use of proteins occurs to the candidate.

*e* **This essay is a mix of some quite good A-level biology and some poor GCSE biology. There are some errors. Most of the essay is relevant, but there are two major digressions. The candidate has given some detail about the structure of a protein molecule, but could have given more. Although the candidate has thought of a reasonable number of protein functions, none is explained in enough detail. For example, the candidate could have made more of the importance of the active site of an enzyme in catalysing reactions.**

**The quality of language is not good. Although the essay is quite easy to follow, much of the language is non-technical and the structure of the sentences is simplistic and clumsy. A number of the biological terms used correctly occur in the 'irrelevant' sections, and so cannot be credited.**

**Likely marks for this essay would be:**
- **Content — 8**
- **Breadth — 3**
- **Relevance — 2**
- **Quality of language — 1**

**The score is 14 out of 25 marks.**

**This is a reasonable essay mark — but it could easily have been better. Leaving out the irrelevant sections would have gained a mark because the whole content would then have been relevant. Taking more care with the construction of the essay would probably have gained another mark. The score could have been 16 marks, without the candidate knowing any more biology.**

# A2 Human Biology

## Candidate B

**(b)** When energy is transferred, some of the energy is lost as heat. The rest of the energy can do useful work, like causing reactions to take place or more physical events.

*e* This is a physical rather than a biological explanation, but it is still relevant.

Some biologists believe that all life depends on the sun as a source of energy, but there are some specialised bacteria that can obtain energy in other ways. The sun's energy is transduced into chemical energy in photosynthesis. In the light-dependent stage, light energy is absorbed by chlorophyll in the grana of a chloroplast and raises the energy level of an electron. The excited electron leaves the chlorophyll and is passed along an electron transport chain, leading to the formation of ATP. Energy is now 'locked up' in the ATP molecule. This is used to drive the light-independent stage where RuBP reacts with carbon dioxide to form two molecules of GP and then TP. Some TP gets recycled to RuBP. The rest is used to make glucose and starch. Starch is stored in storage organs like potatoes.

*e* This is a good description of photosynthesis. The candidate manages to include all the main points without overburdening the essay with too much detail.

When animals eat plants, they obtain the stored food materials. They digest them and absorb the products of digestion. Starch is digested into glucose, proteins are digested to amino acids and lipids are digested to fatty acids and glycerol. Hydrolytic enzymes carry out the digestive processes. The absorbed products can be used for a number of purposes, but some of them are respired. Glucose is the main respiratory substrate. In respiration, the glucose is first converted to acetyl-coenzyme A in glycolysis. Glycolysis has a net yield of 2 ATP molecules. The acetyl-coenzyme A is then converted to pyruvic acid in the link reaction and enters the mitochondrion. Here it goes through the Krebs cycle and then electrons pass down the electron transport chain. A total of 34 ATP molecules are generated from just one glucose molecule.

*e* This is a rather sketchy explanation of respiration and contains a rare error. The candidate has confused acetylcoenzyme A with pyruvic acid.

ATP is the 'energy currency of the cell'. It is a small molecule that can move around cells easily but cannot leave them. It is hydrolysed in a one-step reaction to give ADP, inorganic phosphate and small amounts of energy, which can be used to drive processes within cells. ATP is needed for active transport, biosynthesis of macromolecules, the sodium–potassium ion pump in nerve axons, muscle contraction and many other processes.

*e* This is a *key topic* in this essay, which many candidates do not include.

Energy is transferred between organisms when one organism feeds on another and obtains organic molecules, which it can respire to give ATP. In food chains, not all the energy contained in an organism is passed to the next organism in the chain. This is because not all the organism is eaten and the uneaten parts contain some energy. Also, some of the energy an organism has is used by it. Glucose or fats are respired to produce ATP, but much energy is wasted as heat and is lost to the environment. Also, the parts of the organism that are not eaten die and so some energy passes to decomposers that feed from the dead remains. All the energy that passes into an organism in a food chain must pass out of it either as heat or to the next organism in the chain or to the decomposers.

> *e* This is quite a good account, but there is some confusion between matter and energy.

You don't often get food chains that have more than four links because only 10% of the energy is passed from one link to the next. This means that the fifth organism only has 0.01% of the energy that was present in the first. There isn't enough for another link. This loss of energy can be shown in an ecological pyramid of energy. The food chain, grass ⟶ mouse ⟶ owl would be drawn as:

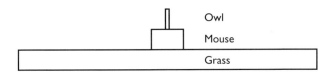

You can see that the amount of energy in each level gets less. This is because of the energy losses to the environment and to the decomposers that can't get passed on up the food chain.

> *e* It is quite legitimate to include one or more diagrams in your essay. Just make sure that the balance is right and it remains an illustrated essay and does not become an annotated flowchart.

Besides respiring aerobically, some organisms respire anaerobically to make ATP. Yeast makes ethanol when it ferments sugars. Humans make lactic acid when they respire anaerobically. Only a small amount of energy is released.

> *e* This is clearly an afterthought and would have been better linked to the earlier section on respiration.

No cell can be alive without the means to release energy to form ATP in order to drive the processes within that cell. This usually means that the cell must be able to respire some organic molecule to release energy. Animals obtain organic molecules by feeding, but they are manufactured by plants in the process of photosynthesis, which transduces light energy into chemical energy.

> *e* This is an excellent concluding paragraph. It draws together all the main themes dealt with in the essay. This is always a good idea, if you have time.

> *e* **This is clearly a much better essay than that of Candidate A, even though they are not writing about the same topic. It is written at a much higher level; there is a lot of detailed A-level biology in this essay. Some areas could have been dealt with in more detail, but examiners recognise that a perfect essay is not**

possible in the time available. This essay covers all the main areas that would be expected. It is presented logically and scientific terminology is used throughout. There are very few errors and no irrelevancies. This essay would probably receive the following marks:
- Content — 12
- Breadth — 3
- Relevance — 3
- Quality of language — 3

The score is 21 out of 25 marks, which is a very good essay mark.

## Examiner's overview

There are 55 marks available for these questions, which represent a synoptic test. Candidate B scores 24 out of 30 marks on the first two questions. A score of 21 marks for the essay gives a total of 45 out of 55 marks — more than enough for a grade A. Candidate A scores only 10 out of 30 marks on the first two questions. However, a score of 14 marks for the essay gives a total of 24 out of 55 marks. This is probably enough to pass.

Without knowing any extra biology, Candidate A could have scored 7 more marks just by:
- not including irrelevant material in the essay and by taking a little more care with language in the essay (2 marks lost)
- reading question 1(a)(ii) more carefully — the candidate clearly knew how to answer it (1 mark lost)
- using language more carefully in question 1(c)(i) — the candidate used the phrase *matching* bases, which could mean either complementary or identical (1 mark lost)
- taking more care with terminology in question 2(c)(i) (1 mark lost)
- supplying a little more detail in question 2(d)(ii), when it was obvious that he/she understood the problem (2 marks lost)

This would have given a total of 31 out of 55 marks — enough for a grade D rather than a borderline E. There were also several instances where just a little more detail would have scored marks. The candidate could probably have achieved a grade C without too much extra effort.

You need to be well prepared for the synoptic paper. To score well, you need facts at your fingertips, which can only happen if you prepare thoroughly. Then:
- read the questions carefully
- don't use vague, imprecise or ambiguous language
- check your answers
- don't include irrelevant material in your essay